Practical Electronics
Second edition

Barry G. Woollard M Phil, C Eng, MIERE, M Inst MC
Technical Director, Beal-Davis Electronics Limited
(Formerly Senior Lecturer in Instrumentation and Control Engineering, and Industrial Electronics, Walsall College of Technology, West Midlands)

D1332127

McGRAW-HILL Book Company (UK) Limited

London · New York · St Louis · San Francisco · Auckland
Bogota · Guatemala · Hamburg · Johannesburg · Lisbon · Madrid
Mexico · Montreal · New Delhi · Panama · Paris · San Juan
São Paulo · Singapore · Sydney · Tokyo · Toronto

Published by
McGRAW-HILL Book Company (UK) Limited
MAIDENHEAD · BERKSHIRE · ENGLAND

British Library Cataloguing in Publication Data

Woollard, Barry G.
Practical electronics. – 2nd ed.
1. Electronic apparatus and appliances
I. Title
621.381 TK7870

ISBN 0-07-084851-3

Library of Congress Cataloging in Publication Data

Woollard, Barry G.
Practical electronics.

Includes index.
1. Industrial electronics. I. Title.
TK7881.W66 1984 621.381 84-3924
ISBN 0-07-084851-3

621.381
WOO

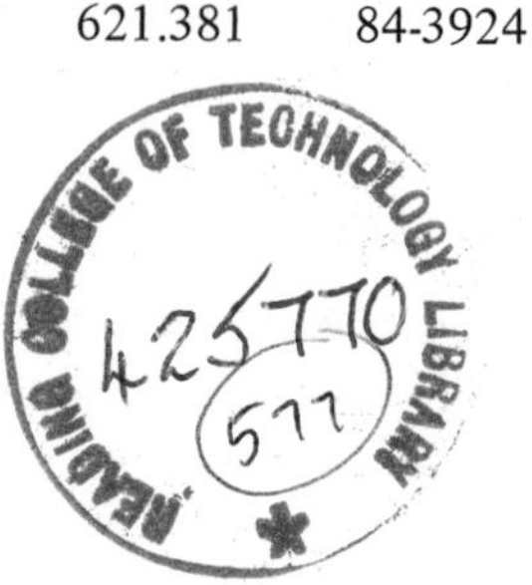

Copyright © 1984 McGraw-Hill Book Company (UK) Limited. All rights reserved. No part of this publication may be reproduced, stored in a retrieval system, or transmitted, in any form or by any means, electronic, mechanical, photocopying, recording, or otherwise, without the prior permission of McGraw-Hill Book Company (UK) Limited, or of the original copyright holder.

1 2 3 4 5 CUP 8654

Typeset in Univers and Press Roman by
STYLESET LIMITED
Salisbury · Wiltshire

Printed in Great Britain at the
University Press, Cambridge

Contents

ACKNOWLEDGEMENT

The photograph on the cover is reproduced by courtesy of Ferranti Electronics Limited.

Preface

With the introduction of electronic devices which complement modern industrial equipment, the electrician is expected to effect repairs in a sphere in which he has little or no experience. The impact of solid-state devices and the advent of integrated circuitry have caused major changes in the physical appearance, size, speed of operation, and capabilities of modern electronic equipment. This book gives an introduction to the components used in industrial electronics with regard to their recognition, ratings, associated circuitry, and typical applications and is developed to include several commonly encountered integrated circuits (ICs) – both linear and digital.

The text has been written with a 'hands-on' development programme in mind, since to gain a *working knowledge* of industrial electronics, it is considered that *practical experience is essential.* In support of this, the descriptive principles have been curtailed to the 'practical' minimum, and many practical exercises have been incorporated to enable the principles and techniques of the components and their associated circuits to be more easily understood. The *Electronic Learning Kit* as shown in the Frontispiece, has been designed and manufactured by *Beal-Davis Electronics Ltd*, Newtown Road, Worcester, as a low-cost means on which all the practical exercises may be performed. The components are mounted on supporting carriers so that the actual components are clearly visible to the user, and labels are used to enable component identification, values, and ratings to be easily determined. These components can be mounted in any position on the matrix board, but it is suggested that components are so mounted that the circuit layout on the matrix board is a 'copy' of the circuit diagram in the text.

Originally developed as a training course, 'Electronics for Electricians', this text has proved useful to a very wide range of people. This, the second edition, has been extended to include a more useful range of components, techniques, and exercises, and is therefore suitable to cover much of the syllabuses for City and Guilds courses and TEC courses, as well as for practising craftsmen, technicians, and engineers.

I gratefully acknowledge all those who have contributed to the realization of this book. In particular, a special thanks to my wife for her patience and efforts in typing the manuscript.

Barry G. Woollard
Longdon, Staffs.

1. Instrumentation

1.1 Introduction

Before any serious work, especially of a practical nature, can be done in electronics it is absolutely essential that a working knowledge of the instrumentation to be used is obtained. In all branches of engineering it is the measurements which will make or break the success of any task. The field of instrumentation alone is vast and to cover this properly would lead to a long course in its own right. What we shall attempt at this stage is to familiarize you with the instrumentation you will need throughout this book, working through some problems and practical exercises to increase your knowledge and also warning you of some of the pitfalls.

One of the basic principles of *all* instrumentation is that *it should not interfere with the system or variable being measured*. To satisfy this basic principle we would need *ideal* instrumentation which was perfect in every way. Such instrumentation does not exist, but some instruments are better than others for a particular task. Others are plain useless and even dangerous if we use them where they are not really suited.

There is a basic need for everyone using instruments to know the range and limitations of the instruments they use.

1.2 Ammeters

Remember that current is a flow of electric charge. An *ammeter* measures the rate of this flow. Hence the current must flow *through* the ammeter. To measure current *break the circuit and insert the ammeter in the break* as shown in Fig. 1.1.

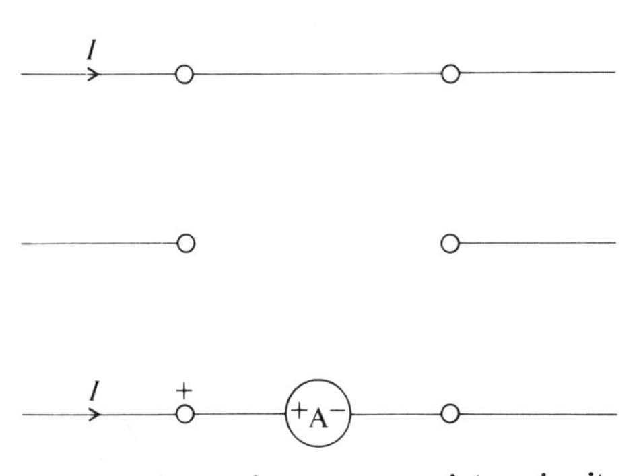

Fig. 1.1 Connecting an ammeter into a circuit.

Also note that for d.c. measurements the current must flow *into* the ammeter positive terminal, therefore it is essential that the meter is connected the right way round, i.e., *positive terminal of the ammeter must be connected to the positive side of the break in the circuit.*

Ammeters may be used to measure a wide range of currents by the use of *shunts*. These are arranged to bypass a definite proportion of the circuit current around the ammeter movement.

Ideally, the ammeter resistance should be zero so that it will not modify the current flow in the circuit in which it is connected. In practice the resistance may vary from thousandths of an ohm to thousands of ohms depending on what it is asked to measure.

If a circuit is supplied by *volts* and has a current of *amperes* flowing in it then the circuit resistance will be of the order of *ohms*, e.g., a 12 V car battery supply to a 50 W headlamp will cause a current of about 4 A to flow:

$$I = \frac{P}{V} = \frac{50}{12} = 4.17 \text{ A}.$$

The headlamp resistance is of the order of ohms:

$$R = \frac{V}{I} = \frac{12}{4.17} = 2.88\ \Omega.$$

An ammeter of resistance 1 Ω would increase the circuit resistance to 3.88 Ω and, therefore, reduce the current, i.e., *the ammeter resistance must be much lower than the circuit resistance.*

If a circuit is supplied by *volts* and has a current of *microamperes* flowing in it then the circuit resistance will be of the order of *megohms*. An ammeter of resistance 1 Ω would make no difference to the circuit resistance in this case. In fact, a *thousand ohms* would not make an appreciable difference. In general:

Ammeters may have resistances of *hundredths of an ohm*.
Milliammeters may have resistances of the order of *ohms*.
Microammeters may have resistances of *thousands of ohms*.

1.3 Voltmeters

A voltmeter is used to measure the e.m.f. produced by an electrical source or the potential difference across two points in a circuit. In all cases voltages exist between two points, i.e., it is the *difference in voltage of one point with respect to another point* that we are measuring. Therefore we simply *prod* the voltmeter connections *across* the voltage we wish to measure, as shown in Fig. 1.2.

Ideally, the voltmeter resistance should be infinite, so as not to interfere with the circuit, i.e., the voltmeter should draw zero current from the circuit. In practice, however, voltmeter resistances vary considerably.

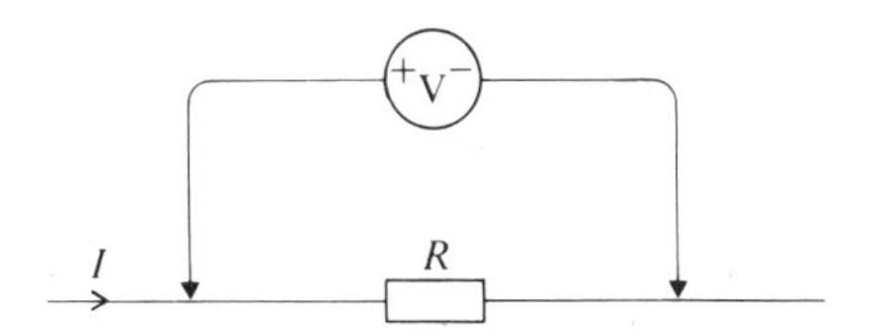

Fig. 1.2 Measurement of voltage.

The *sensitivity* of a voltmeter gives us the necessary information:

$$\text{Sensitivity} = \frac{\text{resistance of voltmeter (ohms)}}{\text{full scale deflection voltage (volts)}}.$$

A typical sensitivity of a d.c. voltmeter, which is a compromise between robustness and low current requirements is 20 000 ohms per volt i.e., a 10 V voltmeter would have a resistance of $R_V = 20\,000 \times 10 = 200\,000\ \Omega$.

Voltmeters can have their range of reading extended by the use of *multipliers*. The multiplier is a resistance connected in series with the meter which limits the current drawn to the same maximum value for a higher voltage range.

1.4 Ohmmeters

An ohmmeter measures the resistance of a circuit or component. Before the resistance can be measured *disconnect the circuit from the power supply* to avoid damage to the ohmmeter, and *disconnect the component from other parts of the circuit* to prevent false readings via parallel resistance paths.

A simple ohmmeter uses a dry battery supply to drive a current through a milliammeter or microammeter. The current is inversely proportional to the resistance to be measured. A variable resistance allows for changes in battery voltage and adjustment of zero resistance indication when the test leads are shorted together. A fixed resistor connected in series limits the current to a pre-determined maximum in case the variable resistor is reduced to zero.

Practical Exercise 1a

The simple ohmmeter

Connect up the circuit arrangement of a simple ohmmeter as shown in Fig. 1.3.

1. Short out the Ω +ve and Ω −ve terminals and adjust the 25 kΩ variable resistor to give 100 μA full-scale deflection (f.s.d.), i.e., this corresponds to *zero* ohms resistance between the terminals.
2. Connect various *known values of resistance in the position* R_x, and note the decrease in reading for increased resistance.
3. Record each meter reading and its corresponding resistance.
4. Is the scale linear?

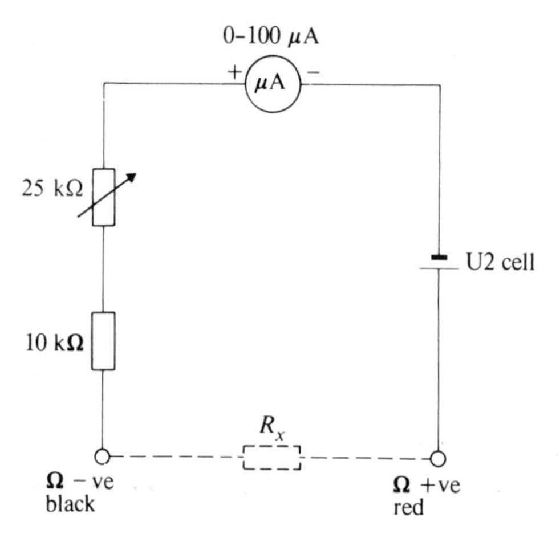

Fig. 1.3 Simple ohmmeter circuit.

5. Now use this arrangement to measure the resistance of a number of unknown resistors.

1.5 Multimeters

Multimeters of the analogue type have been around for quite a long time. The 'AVO' (strictly it should be referred to as the *AVO Multimeter Model 8*) is probably one familiar to you. This instrument uses one meter movement and, by a pair of rotary selector switches, enables the measurement of currents, voltages, and resistances on a wide variety of ranges to be made. Many different models are available and the latest ones conform to the European specification that ranges should always be in multiples of 3 or 10.

The current and resistance ranges are usually quite satisfactory for the type of measurements undertaken, but, as with all instruments of its type, the sensitivity is too low for making measurements in high-impedance–low-voltage semiconductor circuits of modern electronics.

Practical Exercise 1b

Loading effect of the voltmeter

Connect up the circuit shown in Fig. 1.4 to examine the loading effect of the voltmeter.

1. Measure the voltage V_{BC} in the circuit when the two resistors are both 1 kΩ, 100 kΩ, and 1 MΩ, using the multimeter set to 10 V d.c.
   (a) Theoretical value of V_{BC} = __________V, i.e., voltage V_{BC} before voltmeter is connected.

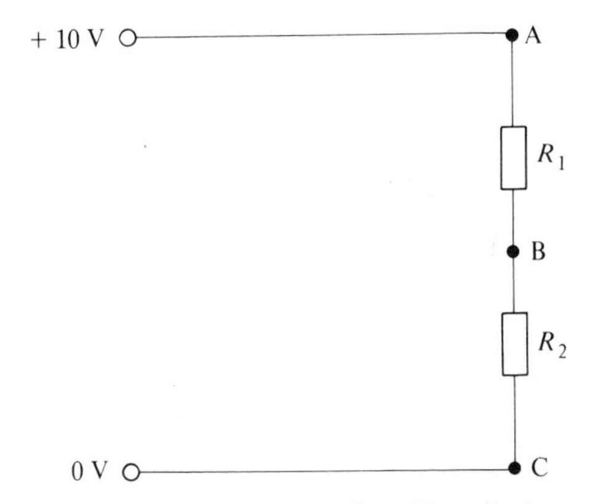

Fig. 1.4 Circuit to examine loading effect of voltmeter.

(b) When the resistors are both 1 kΩ, V_{BC} = __________ V.
(c) When the resistors are both 100 kΩ, V_{BC} = __________ V.
(d) When the resistors are both 1 MΩ, V_{BC} = __________ V.

On the 10 V range, the resistance of the meter is:

$R_v = 10 \times 20\,000\ \Omega = 200\ \text{k}\Omega$.

This is effectively another resistance in parallel with BC, as shown in Fig. 1.5 Now, 200 kΩ in parallel with 1 kΩ makes very little difference, but in parallel with 100 kΩ the resistance between B and C is reduced to:

$$R = \frac{R_2 R_v}{R_2 + R_v} = \frac{100 \times 200}{100 + 200} = \frac{20\,000}{300} = 66.7\ \text{k}\Omega.$$

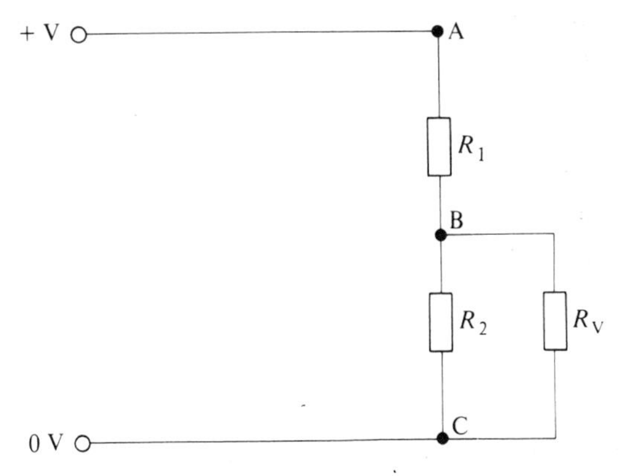

Fig. 1.5 Equivalent circuit of voltage measurement.

Hence the reduced voltage reading. The loading is even more severe when the two resistors are both 1 MΩ.

To measure a.c. or signal voltages we must be aware of the reduced sensitivity – 1000 ohms per volt at best on these ranges. As the voltage range is reduced the current drawn by the meter increases and the sensitivity is drastically reduced. It is impossible to make any low voltage a.c. measurements in normal semiconductor circuits, using an instrument having a sensitivity of only 1000 ohms per volt.

1.6 Digital multimeters

To overcome the above problems, valve voltmeters were developed, but during the last few years these have been largely superseded, as have analogue multimeters, by an extensive range of digital multimeters.

Digital multimeters (DMMs) are relatively more robust. Many have light emitting diode (LED) displays which are frequently up to ¾ in (19 mm) high. An important feature is that they present a constant high impedance on all voltage ranges – typically 10 MΩ.

Many of these instruments work on the dual-slope principle which briefly works as follows: a capacitor is discharged from a fixed level at a predetermined rate to a lower level and then recharges to the original level at a rate proportional to the voltage across the meter terminals. The charge and discharge times are compared and the 'input' voltage calculated and displayed. The current required to operate this circuit is very small and constant, hence the constant 10 MΩ impedance.

Current measurement works by measuring the voltage across a known internal resistance with the unknown current flowing through it. Resistance measurement is made by passing an internally generated constant current through the unknown resistance and measuring the p.d. developed across it.

Overload indication is by flashing display or O/L light.

Reverse polarity causes a POL light to flash, or merely changes the sign of the display. Substantial maltreatment is possible without any damage – but take care anyway.

Some digital multimeters are auto-ranging, i.e., just select voltage, current, or resistance and the instrument chooses the best range and displays the reading.

Accuracy is much improved over general analogue types, being less than 1 per cent and very often 0.1 per cent. Reading error is eliminated by the digital display.

While digital instruments are certainly more easy and clear to read for most people, this can only be true when the quantity being measured is static. To make measurements of quantities which are changing relatively slowly, the analogue meter is the most suitable. Similarly, the analogue meter is more suitable for displaying *trends* in measurements.

Practical Exercise 1c

Comparison of loading effect of digital and analogue voltmeters

Connect up the same circuit used in Practical Exercise 1b shown in Fig. 1.4, and repeat the measurements using both the analogue voltmeter and the digital multimeter (DMM).

	Resistance values		
	1 kΩ	100 kΩ	1 MΩ
Theoretical value of V_{BC}			
Analogue voltmeter reading of V_{BC}			
Digital voltmeter reading of V_{BC}			

Which voltmeter is the best? Why?

1.7 Oscilloscopes

Instruments such as the AVO and many DMMs are calibrated to read steady d.c. or sinusoidal a.c. voltages. Many different waveforms are found in electronic circuits such as rectifier outputs, pulse trains, triangular and rectangular waves, and thyristor waveforms. Readings obtained using conventional instruments are nonsensical and we look for an alternative method of making measurements.

The *cathode ray oscilloscope (CRO)* offers such an alternative by displaying a *picture* of the variable being measured in the form of a graph to a base of time. A beam of electrons, originating from the cathode of the tube, is guided towards the phosphor screen which glows when struck by the electron beam, causing a visible spot. The beam – and hence the spot – is *deflected horizontally at a constant rate* by a voltage generated in the *time-base circuit*, and *vertically* by the voltage of the *incoming signal*. The impedance presented by the instrument is reasonably constant – typically 1 MΩ.

The many controls which are presented on the front of oscilloscopes merely set out to achieve the above requirements. *Do not be baffled by any formidable array of controls – take them one at a time*. The controls can be divided into *three* main groups: *beam* control, *X-deflection control, Y-deflection* control. There are many manufacturers of oscilloscopes and each one has its own approach. Nevertheless, most controls are common to all makes – it is usually the layout that varies. If you know the *function* of all the controls then each time you come across a new model you just have to find each control and set it to the correct position, then, provided the oscilloscope is serviceable, success is assured.

The function of the various controls is outlined as follows:

1. ***Screen.*** This is generally divided into 1 cm squares with the main axes often graduated into 2 mm divisions.
2. ***Scale illumination.*** Variable resistance control of a lamp which highlights the graticule.
3. ***Power-on switch.*** Obvious, but check the indicator – it is pointless carrying on if this does not light.
4. ***Brilliance.*** Controls the intensity of the beam. You can turn this well up on a normal persistence screen. However, take care on oscilloscopes having a long persistence screen, otherwise you will burn the phosphor off the face of the screen.
5. ***Focus.*** Adjusts the spot or trace for a sharp image.
6. ***Astig.*** If this control is present, it will help you to modify an elliptical spot to a round one.
7. ***Y shift.*** Moves the whole trace in the vertical direction.
8. ***Y gain.*** Amplifies or attenuates the incoming signal to obtain the correct vertical size of display. This usually takes the form of a calibrated multi-position range switch, with settings marked in volts per centimetre.

EXAMPLE 1.1

Consider the trace shown in Fig. 1.6, assuming that the *squares* shown have 1 cm sides.

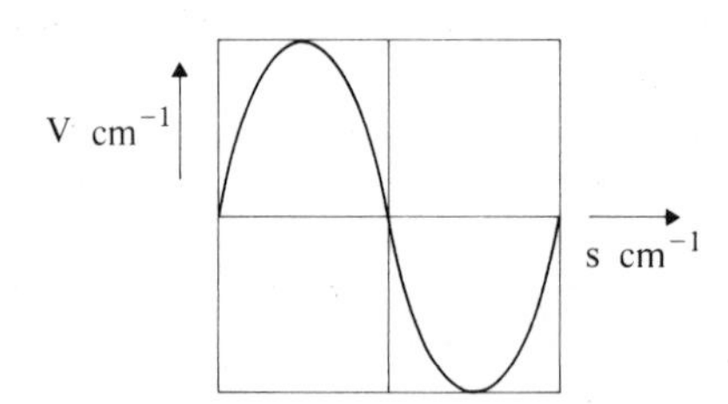

Fig. 1.6 Oscilloscope display.

The trace shown is therefore 2 cm high.

If the Y gain control is set to a sensitivity of 20 V/cm, then the applied signal = 2 x 20 = 40 V, peak-to-peak.

Note: There will be *two* sets of Y controls for a double-beam oscilloscope.

9. ***X shift.*** Moves the whole trace in the horizontal direction.
10. ***X gain.*** Amplifies the X deflection; may be a variable contol and/or PULL x5 or x10. Set to CAL for making measurements on the horizontal axis.
11. ***TB range.*** Time base range switch is calibrated in seconds per centimetre

(s/cm), milliseconds per centimetre (ms/cm), and microseconds per centimetre (μs/cm). This adjusts the speed of the X deflection.

EXAMPLE 1.2

Consider the trace shown in Fig. 1.6 in the horizontal axis.

The trace shown is 2 cm wide.

If the TB range switch is set to a sensitivity of 10 ms/cm, then the applied signal period = 2 x 10 = 20 ms and, the signal frequency = 1/20 ms = 50 Hz.

12. ***TB variable.*** Continuous adjustment between each range setting.
    Note: The range setting is only correct if the variable is on its calibration (or CAL) position.
13. ***Trigger controls.*** These allow the time base to be synchronized to the applied signal to enable a steady trace to be displayed. It is usually a case of either/or selection:
    (a) ***Y_1 or Y_2 or External.*** Allows you to synchronize to the signal fed in to the Y_1 or Y_2 inputs, or to a separate external signal.
    (b) ***+ve or –ve.*** Selects the starting point of the trace.
    (c) ***AC, HF, or TVF/L.*** Normally set to AC, set to HF for high frequencies only, and TVF/L for television servicing.
    (d) ***Auto or Trig. Level.*** When set to AUTO the trace is automatically synchronized to either the +ve or –ve edge. When set to TRIG. Level, this allows you to move the starting point up and down the +ve or –ve slope of the trace.
14. ***AC/GND/DC.*** Select DC for all signals and AC to block out the d.c. levels. The GND position disconnects the signal from the Y amplifier and connects the Y plates to ground (earth). This allows you to preset the Y zero level.
15. ***EXT. socket.*** Provision to feed in the external trigger signal.
16. ***CAL output.*** The output of an internally generated calibration waveform. Usually takes the form of a square wave of frequency 50 Hz and either 0.5 V or 1.0 V peak-to-peak. Allows a quick check on the accuracy of the time base deflecting voltage.
17. ***Chop/Alt.*** A means of getting two separate Y_1 and Y_2 traces from one beam. *Chop selection* causes the beam to change from Y_1 to Y_2 at a very high frequency – typically 250 kHz – and is used on slow TB ranges. *Alt. selection* causes the beam to do a complete Y_1 trace then a complete Y_2 trace, etc. – used on fast TB ranges.
18. ***Trace locate.*** When pressed this will bring the trace to the screen from whereever it is – thus giving an indication which shift controls you need to adjust.
19. ***X–Y.*** Usually a position on the TB range switch. This switches the timebase off and allows signals to be fed into both X and Y inputs to give Lissajous figures and traces.

Practical Exercise 1d

Oscilloscope familiarization

Consider the layout of the oscilloscope shown in Fig. 1.7.

1. Set power switch to ON, and check that indicator lights. Allow a few minutes for the instrument to warm up.
2. Set brilliance (intensity) and focus controls to about mid-position. Set TB range switch to EXT. Set X and Y shift controls to about mid-position. Set X gain and Y gain to CAL position.
3. If the spot is *not* visible, press the *beam locate* button, and adjust the shift controls to bring the spot to the centre of the screen.
4. Adjust the brilliance and focus controls to give a clearly visible sharply defined spot at the centre of the screen.
5. Set the Y amplifier sensitivity to 100 mV/cm, and touch the vertical input terminal. Note that the spot is now deflected to a vertical line due to stray signal pickup by the hand. Set the Y amplifier sensitivity to 2 V/cm and repeat, noting that the vertical deflection is reduced.
6. Set the TB range switch to 10 ms/cm and note that the spot is now deflected to a horizontal line. Adjust the X shift control to ensure that the line extends right across the screen.
7. Select DC and apply a d.c. voltage of about 10 V to the Y input.

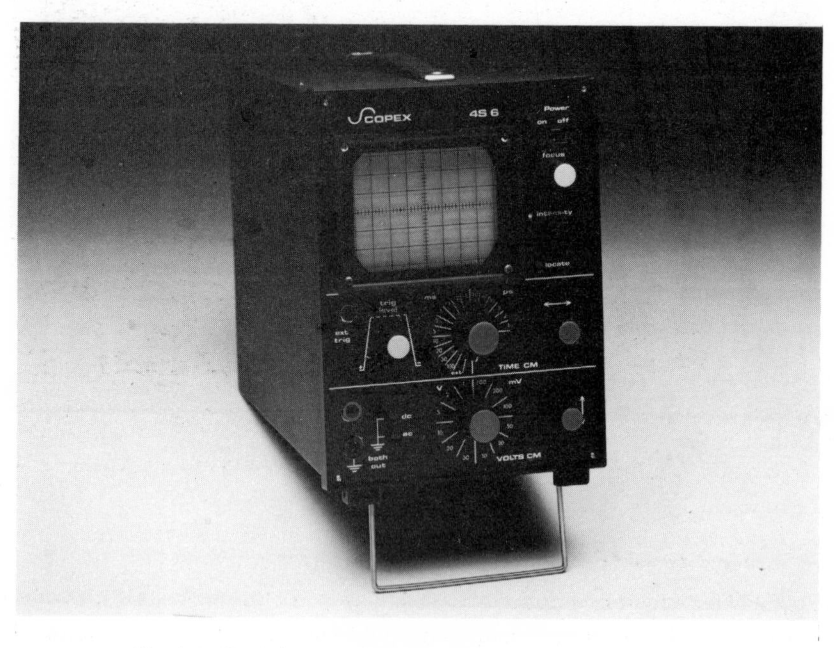

Fig. 1.7 Single beam oscilloscope (*Courtesy of Scopex Ltd*).

Note: It may be necessary to adjust the Y shift to lower the zero voltage level.
Switch from DC to GND and adjust the Y shift to ensure that the trace lies on a graticule mark. Switch back to DC and observe the new position of the line. Determine the d.c. voltage applied to the Y input.

8 Set up the circuit shown in Fig. 1.4 and measure the d.c. voltage V_{BC} using the oscilloscope as described above.
(a) When resistors are both 1 kΩ, V_{BC} = ________V.
(b) When resistors are both 100 kΩ, V_{BC} = ________V.
(c) When resistors are both 1 MΩ, V_{BC} = ________V.

9. Now connect a sinusoidal a.c. signal of frequency about 1 kHz and peak-to-peak amplitude about 10 V from the function generator to the Y input. Set the TB range switch to 0.5 ms/cm and adjust the trig. level until a stationary trace is obtained on the screen. Measure both the amplitude and frequency of the signal.
A typical dual-trace oscilloscope is shown in Fig. 1.8.

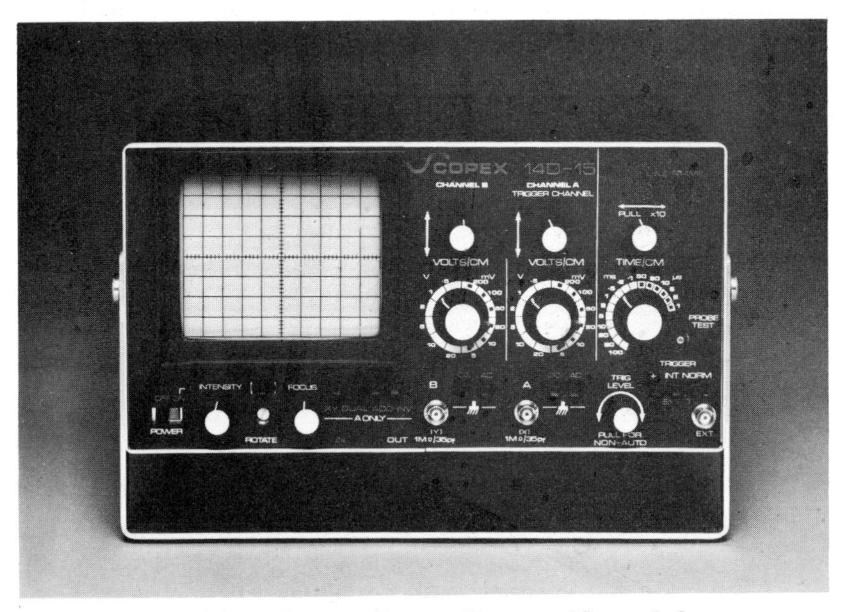

Fig. 1.8 Dual-trace oscilloscope (*Courtesy of Scopex Ltd*).

1.8 Signal generators

The signal (or function) generator is widely used in electronic servicing to feed a waveform of known shape, amplitude, and frequency into a circuit so that it may be traced through the circuit in order to test its performance.

Most of these *function* generators are capable of producing *sine, square, triangular*, and *ramp* waveforms at continuously variable frequencies from fractions of a hertz to several megahertz. Additional features can include a *sweep* facility, which is very useful for frequency response testing and display.

It is always advisable to check with the manufacturer's handbook for any special operating precautions, such as minimum load resistance – typically 50 ohms – that will not overload the generator.

A typical function generator – fitted to the dual-trace oscilloscope – is shown in Fig. 1.9, where the function generator is located along the bottom of the complete instrument.

These instruments are separately powered and switched and can therefore be used independently.

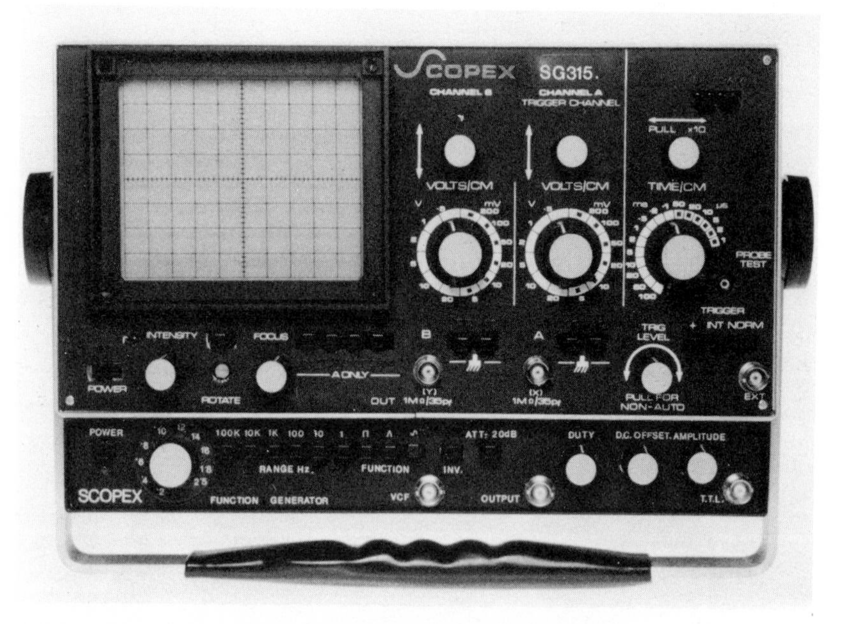

Fig. 1.9 Combined dual-trace oscilloscope and function generator (*Courtesy of Scopex Ltd*).

1.9 Specification of function generator

Typical specification for the function generator shown in Fig. 1.9 is as follows:

1. ***Waveforms.*** Sine, square, or triangular. Selection over the range of 0.2 Hz to 250 kHz is by push button switches and a calibrated variable. The waveshape, polarity, and mains on/off are also selected by push buttons.

2. *D.c. offset variable* – allows positioning of the selected waveform within the output range.
3. *Duty cycle variable* – range of approximately 10 to 1 enabling the generator to become a very useful pulse, ramp, and skewed sine wave generator.
4. *Output:* 20 V peak-to-peak from 600 Ω with a variable control and a fixed 20 dB push-button attenuator.
5. *TTL sync. output* – capable of driving up to 20 TTL loads. This output is convenient to use for triggering the oscilloscope.
6. *VCF input:* Voltage Controlled Frequency input provides the means by which an external voltage can sweep the generator over its frequency range.

1.10 Specification for the dual-trace oscilloscope

Typical specification for the dual-trace oscilloscope featured in Fig. 1.8 and Fig. 1.9 is as follows:

1. *Vertical system.* (Channels A and B)
   (a) *Sensitivity:* 5 mV/div to 20 V/div (12 calibrated ranges).
   (b) *Accuracy:* ± 3 per cent.
   (c) *Bandwidth:* (–3 dB) : d.c. coupled 15 MHz
   a.c. coupled 3 Hz to 15 MHz.
   (d) *Risetime:* approx 24 ns
   (e) *Max input voltage:* 400 V (d.c. + peak a.c. to 10 kHz)
   (f) *Input impedance:* 1 MΩ ± 3 per cent and 33 pF (approx)
   (g) *Operating Modes:* Dual trace alternate or chop (110 kHz), channel A only; X – Y; Add; Invert channel B.
   (h) *X–Y Operation:* Channel A is switched to the (X) horizontal deflection system, channel B remains the (Y) vertical amplifier. Bandwidth d.c. to 1 MHz. Phase shift $< 3^\circ$ at 50 kHz.
2. *Horizontal system* (time base)
   (a) *Sweep speeds:* 1 μs/div to 100 ms/div (16 calibrated ranges).
   (b) *Accuracy:* ± 3 per cent
   (c) *Magnifier:* x 10 (increases fastest sweep to 100 ns/div).
3. *Trigger Circuit*
   (a) *Source:* Internal Channel A
   External.
   (b) *Modes:* Normal TV (line, frame via active sync separator).
   (c) *Sensitivity:* Internal normal 3 mV, 10 Hz–15 MHz.
   Internal TV 5 mV, Sync Pulse Amplitude.
   External normal, 300 mV, 10 Hz–15 MHz.
   External TV 150 mV. sync pulse amp.
   (d) *Max. Input:* 100 V (d.c. and peak a.c. to 1 kHz)
   (e) *Bright line Auto:* trace free runs at sweep range in absence of a trigger signal. Pull switch to disable bright line auto.

(f) ***Level control:*** selects trigger level point.

4. ***Display.***
   (a) ***Graticule:*** 10 x 8 div (1 div = 1 cm)
   (b) ***CRT:*** P31 phosphor standard. P.7 optional.
   (c) ***Trace locate:*** returns overscanned trace to display area.
   (d) ***EHT:*** 2 kV
   (e) ***Z modulation:*** + 5 V for perceptible modulation.
   (f) ***Input socket:*** B N C.
5. ***Probe Compensation.***
   (a) ***Output voltage:*** 3.5 V peak-to-peak (approx)
   (b) ***Waveshape:*** square wave
   (c) ***Frequency:*** at sweep repetition rate.

2. Resistors

2.1 Introduction

Resistors used in industrial electronics fall into two main categories:

1. ***Linear resistors.*** Those which obey Ohm's law.
2. ***Non-linear resistors.*** Consist of three types in common use:
   (a) Photoresistors – light sensitive.
   (b) Thermistors – heat sensitive.
   (c) Voltage-dependent resistors.

2.2 Linear resistors

The circuit symbols for linear resistors are as shown in Fig. 2.1, and the units in which resistance is measured are *ohms* (symbol: Greek capital omega, Ω). Multipliers of 10^3 lead to the common usage of:

kilohms	(kΩ)	1000s
megohms	(MΩ)	1 000 000s

In many circuit diagrams and manufacturer's literature the decimal point is indicated by the position of the multiplier letter.

Fig. 2.1 Circuit symbols for linear resistors.

EXAMPLE 2.1

1. 4700 Ω = 4.7 kΩ = 4K7.
2. 3 300 000 Ω = 3.3 MΩ = 3M3.
3. 6.8 Ω = 6R8.

In addition, a letter system is used to indicate the percentage tolerance:

F = ±1 per cent; G = ±2 per cent: J = ±5 per cent:
K = ±10 per cent; M = ±20 per cent.

EXAMPLE 2.2

Determine the value of the resistor shown in Fig. 2.2.

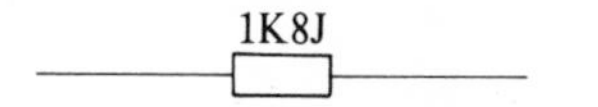

Fig. 2.2 Resistance and tolerance.

Solution

The resistor shown is 1.8 kΩ ± 5 per cent.

Many manufacturers produce a wide range of fixed resistors which may be: wirewound, tape, metal film, metal oxide film, cermet, carbon element.

The percentage tolerance governs the values of the resistors available in a particular range. The nominal values are chosen so that the tolerance ranges usually meet each other or overlap. These chosen nominal values are called *preferred* values.

EXAMPLE 2.3

Consider the ±10 per cent tolerance resistor range:

1. 100 Ω can be 90 to 110 Ω.
2. 120 Ω can be 108 to 132 Ω.

Preferred values are chosen in the range 100 to 1000 and are then manufactured together with all their multiples of ten within the total range. The preferred values in the 10 per cent tolerance range are:

100, 120, 150, 180, 220, 270, 330, 390, 470, 560, 680, 820

2.3 Wattage

The most important fact after the resistance is the *power rating* or wattage of the resistor. For a required resistance the wattage may be calculated using:

$$W = VI = I^2R = V^2/R.$$

The type of resistors used in electronics are rated from $\frac{1}{8}$ W upwards, e.g., $\frac{1}{8}, \frac{1}{4}, \frac{1}{2}$, 1, 2, 5, 10, etc.

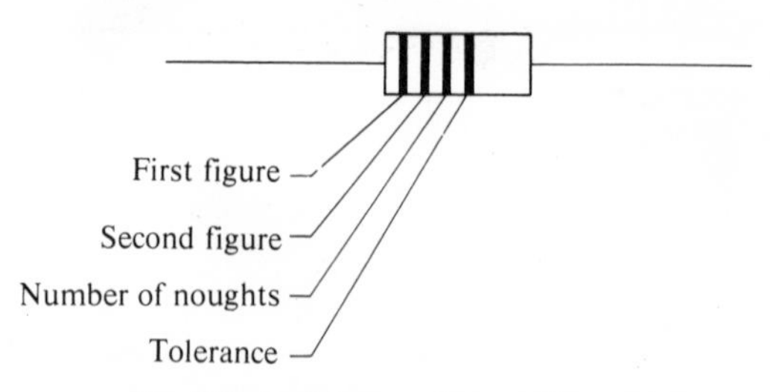

Fig. 2.3 Resistance colour coding system.

2.4 Colour coding

Many small resistors have their value indicated by a colour-coded band system, which may be identified as shown in Fig. 2.3.

The colour code is as shown in Table 2.1.

Table 2.1 Resistor colour code

Colour	Figure	Tolerance (per cent)
Black	0	
Brown	1	±1
Red	2	±2
Orange	3	
Yellow	4	
Green	5	
Blue	6	
Violet	7	
Grey	8	
White	9	
Gold	—	±5
Silver	—	±10
None	—	±20

EXAMPLE 2.4

Identify the value and tolerance of the coded resistor shown in Fig. 2.4.

Solution

Brown 1
Grey 8
Red 2 noughts
Gold ±5 per cent tolerance.
Resistor = 1 K8J = 1.8 kΩ ± 5 per cent.

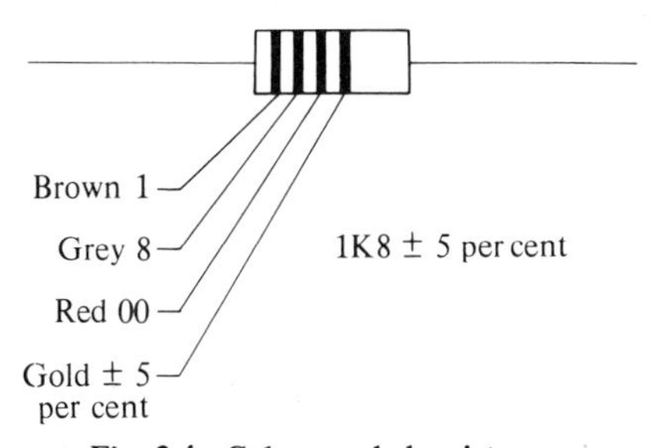

Fig. 2.4 Colour-coded resistor.

On some older resistors, a fifth (salmon pink) band indicated HY STAB, or high stability, in which the resistance was guaranteed not to drift out of the tolerance over a long life. This has now ceased in favour of manufacturer's package marking.

Practical Exercise 2a

Series resistors

Connect up the circuit arrangement show in Fig. 2.5, and adjust the supply voltage to 9 V d.c.

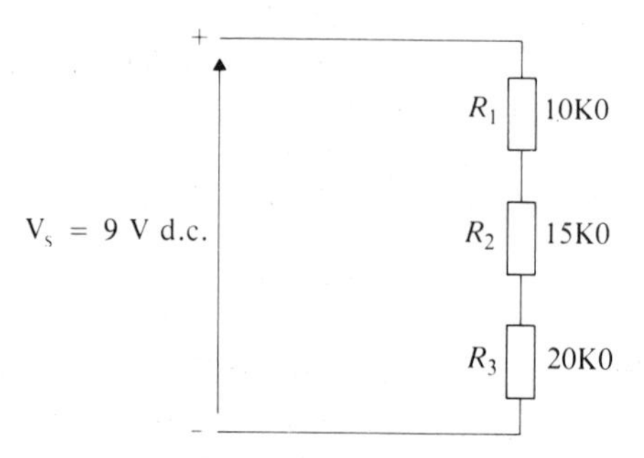

Fig. 2.5 Series resistors.

1. Measure the voltage drop across each resistor, and record the readings.
2. Compare the readings with one another. Why is the voltage across the 20 kΩ resistor greater than the voltage across the 10 kΩ resistor, and by how much is it greater?
3. What is significant about the *sum* of the voltage drops 3.
4. Determine the equivalent circuit resistance R_T, ($R_T = R_1 + R_2 + R_3$).
5. Calculate the circuit current, ($I = V_s/R_T$).
6. Switch off the supply. Connect an ammeter (selected to a suitable range) into the circuit. Switch on and record the ammeter reading.
7. Compare the calculated current with the measured current.

Practical Exercise 2b

Parallel resistors

Connect up the circuit arrangement shown in Fig. 2.6, and adjust the supply voltage to 9 V d.c.

1. Measure the voltage across each resistor. It should be the same!
2. Calculate the current in each resistor.
3. Switch off the supply. Connect an ammeter (selected to a suitable range) in series with R_1. Switch on and record ammeter reading.

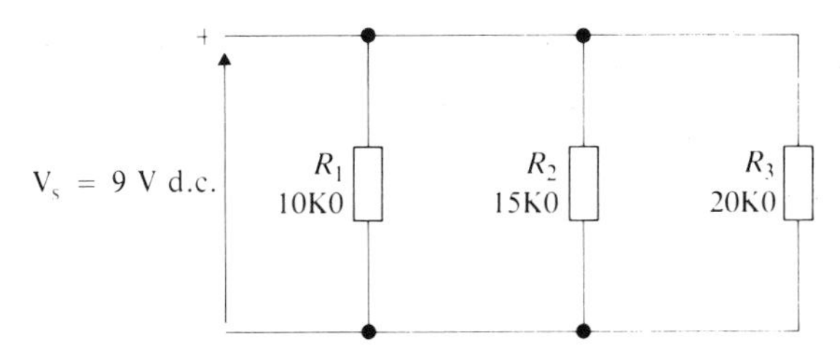

Fig. 2.6 Parallel resistors.

4. Repeat (3) for each resistor in the circuit.
5. Compare the measurements for current with calculated values.
6. Determine the total circuit current, $I_T, = I_1 + I_2 + I_3$.
7. Determine the equivalent circuit resistance, $R_T = V_s/I_T$.
8. Determine the equivalent circuit resistance, R_T as follows:

$$\frac{1}{R_T} = \frac{1}{R_1} + \frac{1}{R_2} + \frac{1}{R_3}$$

Note: A useful technique for parallel connected resistors is to deal with two at a time, e.g., if $R_1 = 10\Omega$, $R_2 = 20\Omega$ and $R_3 = 30\Omega$, the equivalent circuit resistance may be determined by firstly taking R_1 and R_2 in parallel, then the equivalent of that is taken in parallel with R_3:

$$R_p = \frac{R_1 R_2}{R_1 + R_2} = \frac{10 \times 20}{10 + 20} = \frac{200}{30} = 6.67\,\Omega$$

$$\text{and } R_T = \frac{R_p R_3}{R_p + R_3} = \frac{6.67 \times 30}{6.67 + 30} = \frac{200}{36.67} = 5.45\,\Omega.$$

9. Compare the two values of R_T.

Practical Exercise 2c

Series–parallel resistors

Connect up the circuit arrangement shown in Fig. 2.7, and adjust the supply voltage to 9 V d.c.

1. Measure the voltages across the parallel branch, and across the 20K0 series resistor.
2. Determine the equivalent resistance of the parallel branch.
3. Determine the equivalent resistance of the circuit.
4. Calculate the circuit current, $I_T = V_s/R_T$.
5. Connect an ammeter to measure the circuit current. Compare this reading with the calculated value.

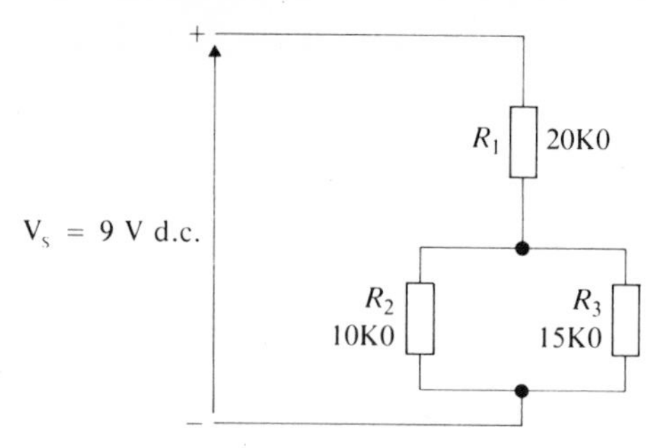

Fig. 2.7 Series-parallel resistors.

2.5 Photoresistors

The circuit symbols for photoresistors are as shown in Fig. 2.8. These are usually based on the fact that a film of cadmium sulphide has a high resistance in the absence of light but has a reduced resistance when light is directed on its surface. The dark resistance may be of the order of megohms and reduce to a few hundred ohms in daylight.

Fig. 2.8 Circuit symbols for photoresistors.

The photoresistor can be used in a potential divider network to cause a voltage to change if the light falling on it changes. This principle can be used in automatic street light switching or parking light control.

Practical Exercise 2d

Photoresistor operation

Connect up the circuit arrangement shown in Fig. 2.9 (*a*) and measure V_{BC} when light is directed on the photoresistor and when the photoresistor is shielded from the light.

1. When light was directed on the photoresistor, the voltage V_{BC}________ from________V to ________V.
2. Repeat this procedure with the circuit arranged as shown in Fig. 2.9 (*b*). When light was directed on the photoresistor the voltage V_{BC}________ from ________V to________V.
3. With the circuit connected as shown in Fig. 2.9 (*b*), shield the photoresistor ORP 12 from daylight and direct it towards a fluorescent light.

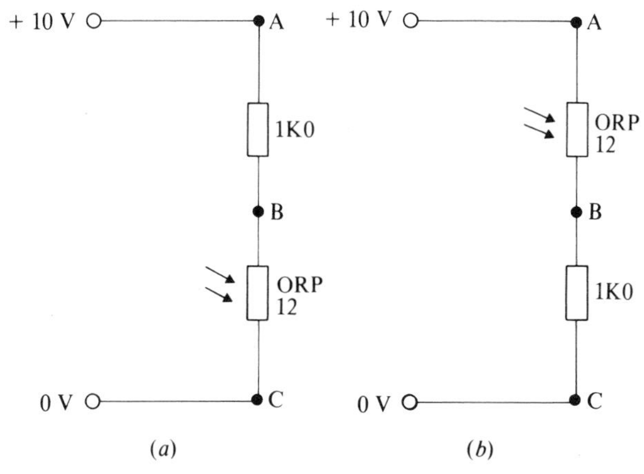

Fig. 2.9 Photoresistor operation.

Connect the oscilloscope across the 1K0 resistor and see if you can pick up the variation of current which should result if the fluorescent tube is going on and off with the a.c. supply.

What frequency is the fluorescent tube flashing at?

This shows that the photoresistor can be used to sense light changes which the human eye is too slow to pick up.

2.6 Thermistors

The circuit symbols for thermistors are as shown in Fig. 2.10.

Thermistors change their resistance with change of temperature in a rather exaggerated way. Two types are commonly available: positive temperature coefficient (p.t.c.) and negative temperature coefficient (n.t.c.).

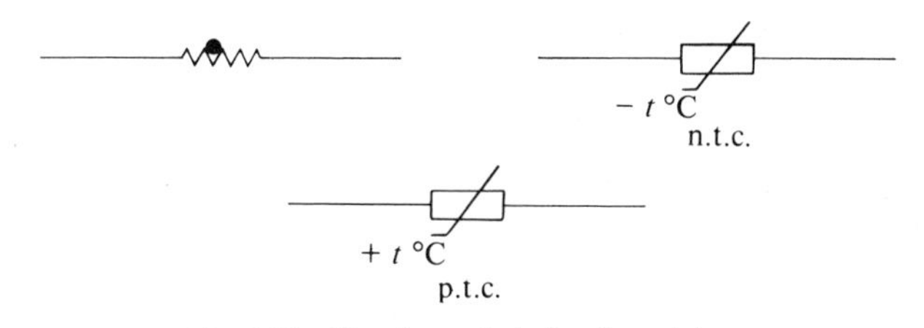

Fig. 2.10 Circuit symbols for thermistors.

N.t.c. thermistors are supplied in many shapes and sizes from beads to larger rods or discs. The range of 'cold' resistance is extensive from a few ohms to several megohms. As the temperature increases due either to self heating or external heating, the resistance decreases, e.g. hundreds of ohms at room temperature to tens of ohms at 100°C. These find application in many ways, some of which include:

1. ***Car temperature gauge.*** The circuit arrangement is as shown in Fig. 2.11 Increased water temperature reduces the resistance, allowing more current to flow, and the meter reading increases.

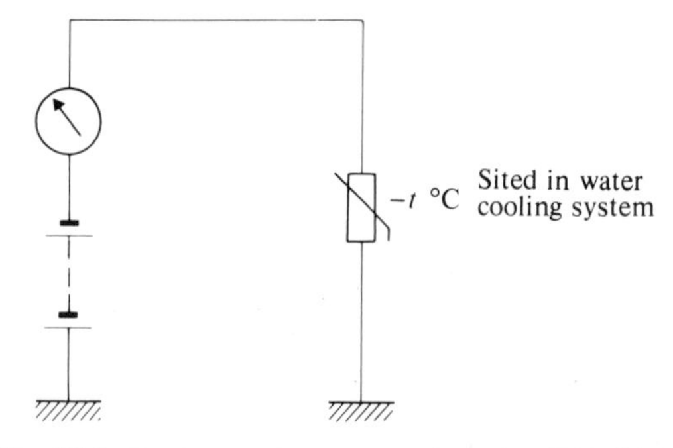

Fig. 2.11 Car temperature gauge using n.t.c. thermistor.

2. ***Spontaneous combustion alarm*** in a refuse tip where coal or metal waste is involved. The increased current trips an alarm when the thermistor, buried in the refuse, senses an increase of temperature which may lead to a spontaneous fire.
3. ***Control of heater current.*** The thermistor is connected in series with the heater, and its resistance decreases due to self heating, allowing the heater current to rise to its normal value at a reduced rate.

P.t.c. thermistors increase their resistance with increase of temperature in a rather dramatic way, as shown in the characteristic in Fig. 2.12. As the temperature increases from room temperature the resistance may decrease slightly, but, at a predetermined reference temperature, the resistance increases at a very rapid rate. This change may be from hundreds of ohms to hundreds of kilohms in a few degrees.

P.t.c.'s find applications in overheating protection systems of machines. Being very small they can be buried in the windings of an electrical machine during manufacture and connected into a safety override system during use. Excessive heating could lead to the tripping of the power supplying the machine. Alternatively, mounted in the bearings of a ship's propellor shaft, they could be used to set off an overheat alarm warning.

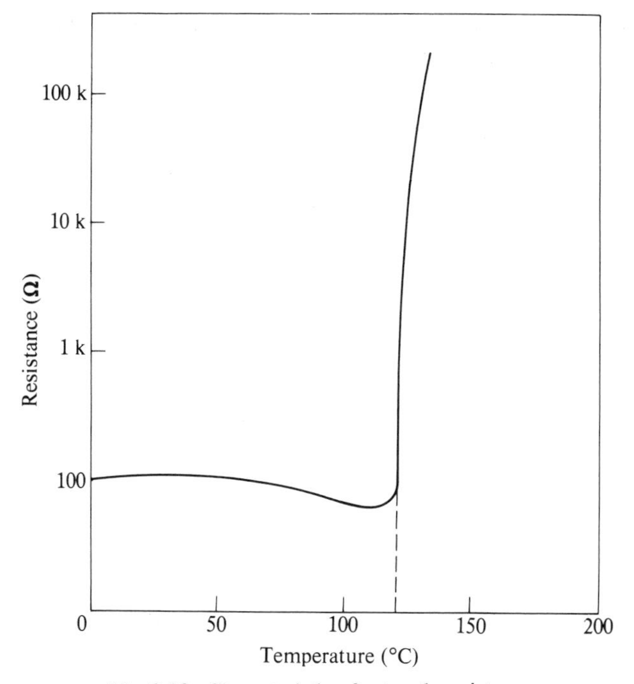

Fig. 2.12 Characteristic of p.t.c. thermistor.

Practical Exercise 2e

N.t.c. thermistor as a current limiter

1. Measure the 'cold' resistance of the 12 V 1 W filament lamp supplied in the Kit.
2. Connect up the circuit shown in Fig. 2.13(*a*), switch on, and note the instantaneous maximum value of the current drawn by the filament lamp.

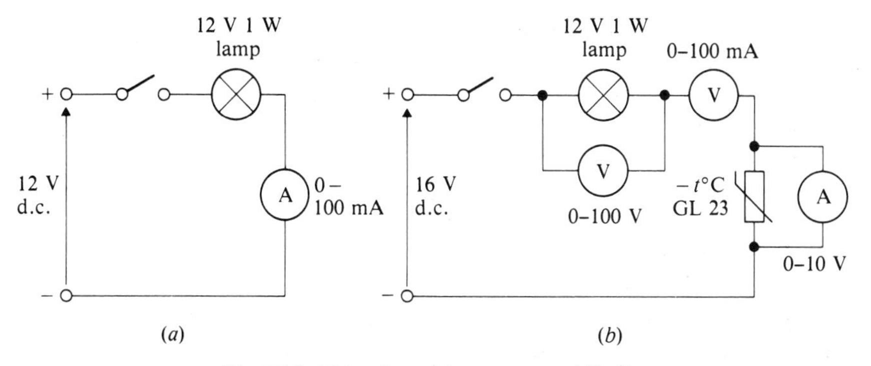

Fig. 2.13 N.t.c. thermistor as a current limiter.

3. Now connect the thermistor and filament lamp in series as shown in Fig. 2.13(*b*). Switch on and note the initial values of the current and voltage drops across lamp and thermistor.
4. Sit back and allow the thermistor time to heat up. After a while, the lamp will have reached its working temperature. Measure the current and the voltage drops.
5. Calculate the 'hot' and 'cold' resistances of the thermistor and filament lamp.

2.7 Voltage-dependent resistors

The circuit symbol of the voltage-dependent resistor (VDR) is as shown in Fig. 2.14. Sometimes called a metrosil, the VDR has a resistance which decreases with increase of applied voltage. This change of resistance is not linear; doubling the voltage can decrease the resistance to one-tenth of its original value.

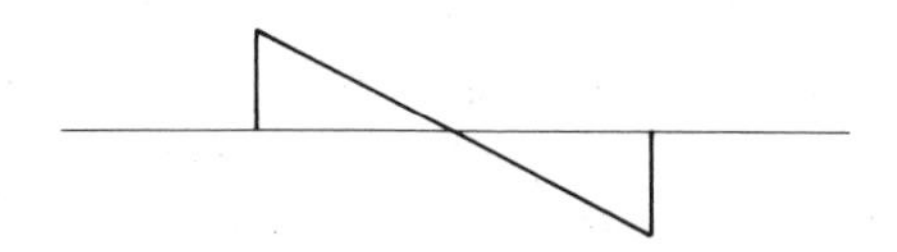

Fig. 2.14 Circuit symbol for a voltage-dependent resistor.

VDRs are used in the suppression of voltage surges to protect other circuits, and they are usually in the form of a disc which is colour coded by the manufacturer to indicate the reference voltage.

3. Capacitors

3.1 Introduction

Capacitors are extensively used in electronic circuits and perform a wide variety of functions. A capacitor is basically a device which stores electric charge. Any two conducting surfaces (plates) separated by an insulating medium form a capacitor. If electrons are drawn from one of the plates and passed to the other, then a state of charge exists between them across the insulating medium, i.e. a positive charge on the plate which has lost electrons and a negative charge on the plate which has gained electrons.

This has been stated in general terms to emphasize the fact that a capacitor can be formed anywhere where the above conditions exist, i.e., capacitors can be made using this idea and 'nuisance' capacitors can also be found in places where we do not want them, e.g., any two pieces of insulated wire running together or a semiconductor device junction, etc.

3.2 Capacitance

Charge (symbol Q) is measured in *coulombs*, and a capacitor which is charged will have a voltage between its terminals of V volts.

The ability of a capacitor to store charge is called its *capacitance* (symbol C) and is measured by the amount of charge that can be stored for a given rise in voltage.

$$\text{Capacitance, } C = \frac{\text{charge, } Q}{\text{voltage, } V}$$

or

$$C = Q/V.$$

The conducting surfaces of a capacitor are usually in flat 'plate' form. The size of the capacitance depends upon the plate area, A, the distance between the plates, d, and the insulating medium. Capacitance may be calculated from the formula

$$C = \epsilon A/d,$$

Where $\epsilon = \epsilon_0 \epsilon_r$; ϵ_0 = permittivity of free-space, a constant; ϵ_r = relative permittivity, a multiplying factor which depends upon the insulating medium, or *dielectric* material used between the plates.

The units in which capacitance is measured are *farads* (symbol F).

3.3 Units

A capacitor is said to have a capacitance of 1 F if a charge of 1 C raises its voltage by 1 V.

However, the farad is too large to be used as a practical unit and it is, therefore, common practice to use the microfarad and other sub-multiples:

$$1 \text{ microfarad} = 1\,\mu\text{F} = 1 \times 10^{-6}\text{ F},$$
$$1 \text{ nanofarad} = 1\,\text{nF} = 1 \times 10^{-9}\text{ F},$$
$$1 \text{ picofarad} = 1\,\text{pF} = 1 \times 10^{-12}\text{ F}.$$

3.4 Voltage rating

This is a most important fact when considering capacitors. It tells you the *maximum* voltage that can be applied between the plates of a capacitor, without breaking down the insulating dielectric.

Note: there is little margin for error, i.e., no 2:1 or greater 'factor of safety' as in civil engineering, so you must observe the maximum limits.

The rating is also a d.c. one, i.e., a 200 V capacitor can be connected across a maximum of 200 V d.c.

For a.c. supplies, the supply voltage stated, e.g. the domestic mains supply 240 V, is the r.m.s. value of voltage, and to determine the voltage rating of a capacitor to be used on this supply we must calculate the maximum (or peak) value of supply voltage:

$$V_{max} = \sqrt{2} \times 240\text{ V}$$

assuming a sine wave supply. Therefore,

$$V_{max} \approx 340\text{ V}$$

and we would, therefore, choose a capacitor having a rating of, say, 400 V for use on the 240 V a.c. mains.

3.5 Stored energy

The energy stored in a capacitor is given by the formula

$$\text{Energy} = \tfrac{1}{2}\,CV^2\text{ joules.}$$

This energy can remain in the capacitor for a considerable time, even when the supply to a circuit has been disconnected. *Take care* when dealing with circuits in which the capacitors are charged up to high voltages during normal operation. These capacitors must be discharged before they can be handled safely.

After the supply has been disconnected these capacitors may be discharged as follows:

1. ***By connecting a resistive link across the terminals.*** A few kilohms will be

low enough to ensure rapid discharge of most capacitors, but not too low to cause a hazard. A short-circuit link may produce a very dangerous flash. Take care to isolate yourself from the link.

2. ***By the use of a bleed resistor.*** This is a resistive link which is permanently connected across large capacitors to discharge them automatically when the supply voltage is removed from the circuit. Their value is usually high, and may be of the order of megohms, so as not to interfere with the normal circuit operation. This means that it may take a considerable time for the capacitor to be completely discharged, e.g., several minutes. Always check, or discharge using a resistive link to be safe.

3.6 Construction and coding

There are many different capacitors in use, but all of them fall into two basic groups: *non-electrolytic* – which are not polarized – and *electrolytic* – which are polarized, i.e., they have positive and negative terminals.

1. ***Non-electrolytic.*** The circuit symbols are as shown in Fig. 3.1. In order to get the largest capacitance into the smallest weight and volume of capacitor, the plates are usually made of a thin silver or aluminium foil. These are usually rolled up and separated by waxed paper, polythene, polyester or polycarbonate films, etc., and enclosed in some form of plastic or metal encapsulation. Where the dielectric will not bend, e.g., mica, the capacitor becomes a flat-plate type. These can be made in a variable form in which a screw arrangement compresses the layers to increase the capacitance. As the screw is tightened or slackened the capacitance increases or decreases, respectively. Capacitors of this type have a maximum of a few hundred picofarads and have replaced many air dielectric variable capacitors for radio tuning purposes.

Fig. 3.1 Circuit symbols for (*a*) fixed and (*b*) variable non-electrolytic capacitors.

In the general range of capacitors the many types may have the same capacitance but one may perform a particular function, such as interference suppression, best. You are advised to consult the manufacturer's data.

Nominal values are produced in ranges of preferred values, similar to those of resistors but not in such a coherent form. Tolerances vary widely from ± 20 per cent down to ± 1 per cent with the cost usually increasing sharply as the tolerance decreases.

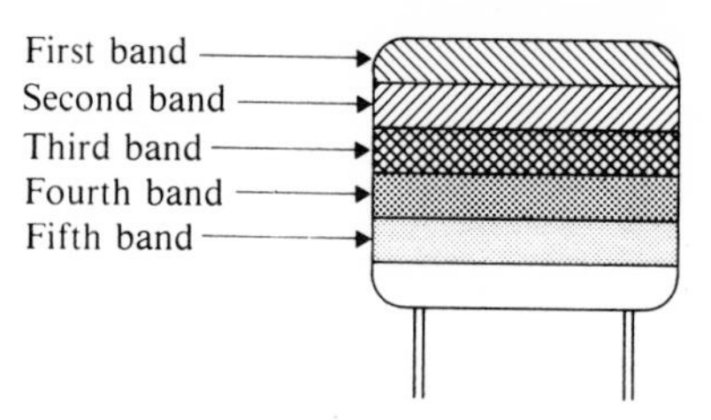

Fig. 3.2 Polyester capacitor colour coding system.

Polyester capacitors may have their value indicated by a colour-coded band system, which may be identified as shown in Fig. 3.2.

The colour-code is as shown in Table 3.1

Table 3.1 Capacitor colour code

Colour	First and Second Bands	Third Band	Fourth Band (Tolerance)	Fifth Band (Working Voltage)
Black	0		± 20%	
Brown	1			
Red	2			250 V d.c.
Orange	3	x 0.001 μF		
Yellow	4	x 0.01 μF		00 V d.c.
Green	5	x 0.1 μF		
Blue	6			
Violet	7			
Grey	8			
White	9		± 10%	

EXAMPLE 3.1

Identify the value, tolerance, and working voltage of the coded polyester capacitor shown in Fig. 3.3.

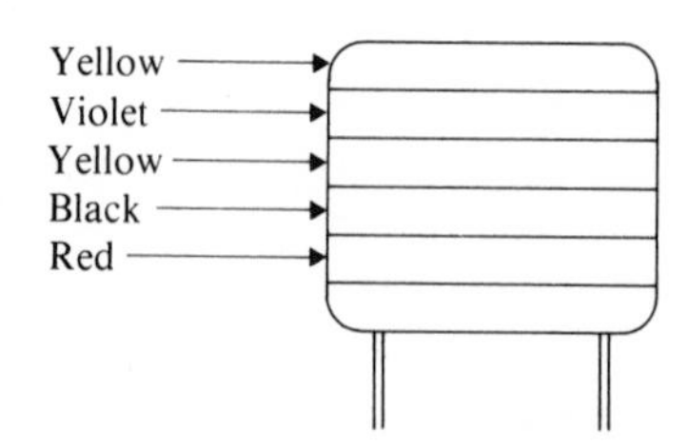

Fig. 3.3 Colour-coded capacitor.

Solution

Yellow	4
Violet	7
Yellow	x 0.01 μF
Black	± 20 per cent
Red	250 V d.c.

Therefore, the value of capacitance = 47 x 0.01 = 0.47 μF
having a tolerance = ± 20 per cent
and a voltage rating = 250 V d.c.

2. ***Electrolytic.*** The circuit symbol is as shown in Fig. 3.4. Early electrolytics were formed by oxidizing one of the aluminium plates and replacing the dielectric by wet *electrolytes* – hence the name. The capacitance is formed at the oxide layer. Paste electrolytes made the production of very large capacitance, small size, foil capacitors possible, i.e., something like 100:1 gain over conventional non-electrolytic types. The main disadvantage is that the capacitance is only obtained one way, i.e., they are *polarized.* This fact is indicated by the different symbol, and it is *extremely important to connect the capacitor terminals to the correct polarity of the supply. Incorrect connection results in a short circuit and consequent destruction of the capacitor.*

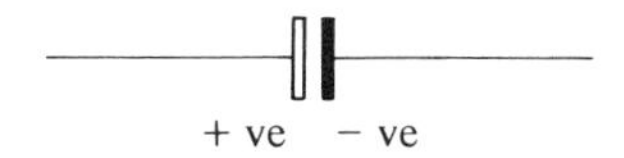

Fig. 3.4 Circuit symbol for electrolytic capacitor.

There are various ways by which the terminals of electrolytic capacitors can be identified:
(a) A positive sign nearer the positive end, as shown in Fig. 3.5(*a*).
(b) A red stopper in the positive end, and a black or blue one in the negative end.

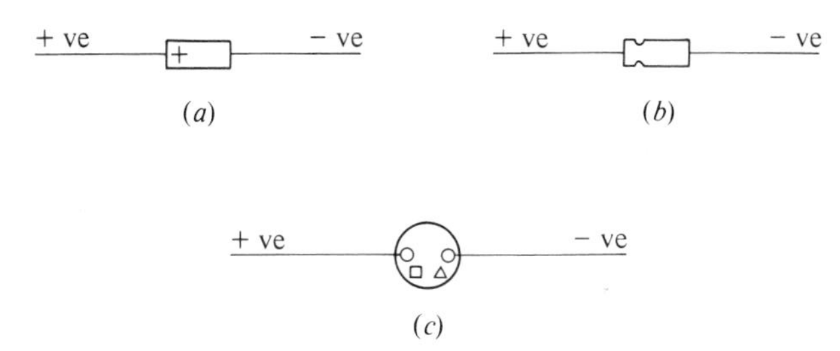

Fig. 3.5 Identification for terminals of electrolytic capacitors.

(c) A swage at the positive end, as shown in Fig. 3.5(*b*).
(d) Terminals stamped as shown in Fig. 3.5(*c*) when the leads come out of the same end.

If in doubt consult the manufacturer's literature.

Note: Aluminium electrolytics will only withstand about 10 to 15 per cent of the rated voltage applied in reverse, without damage.

Another disadvantage is that tolerances on nominal values are high and asymmetrical. Typical ranges are –10 per cent + 50 per cent for good ones and –20 per cent + 100 per cent for cheaper ones. The capacitance also tends to drift upwards with age, which means that they cannot be used where any reasonable degree of stable accuracy is required. However, they do find extensive use in the smoothing of rectified a.c. supplies.

Electrolytic capacitors deteriorate in storage – their leakage currents increase. If they have been stored for more than 12 months they will usually require a *reforming* charge using a circuit such as that shown in Fig. 3.6. They should be connected to a variable d.c. supply through a high resistance and the voltage gradually raised, while *keeping the leakage current below a specified maximum*, until the rated voltage is built up across the capacitor. Again, refer to manufacturer's data.

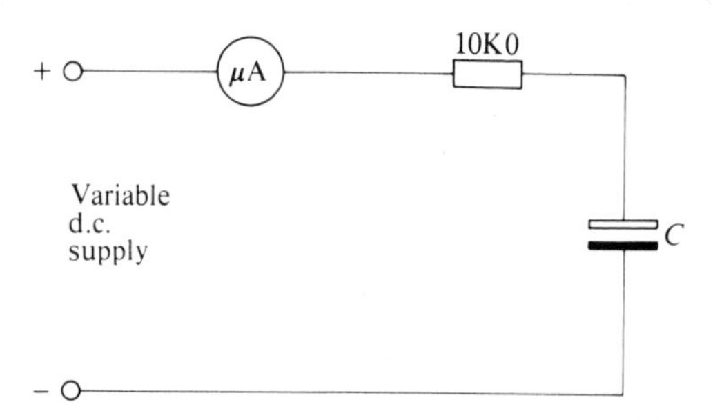

Fig. 3.6 Reforming charge for electrolytic capacitors.

Recent developments have produced a new type of electrolytic, using tantalum and tantalum oxide, which gives further capacitance/size advantages. As with all electrolytics, the voltage ratings available tend to be low, e.g. 25 V, 63 V typically, but the tantalum capacitor is particularly vulnerable to reverse voltages, 0.3 V being a typical maximum reverse voltage. *Watch it, even a 1.5 V U2 cell (ohm-meter on Ω or Ω ÷ 100 range) can blow them!*

Tantalums tend to be small 'blob' types with the two leads protruding from the bottom, the polarity and values may be printed on or coded, as shown in Fig. 3.7.

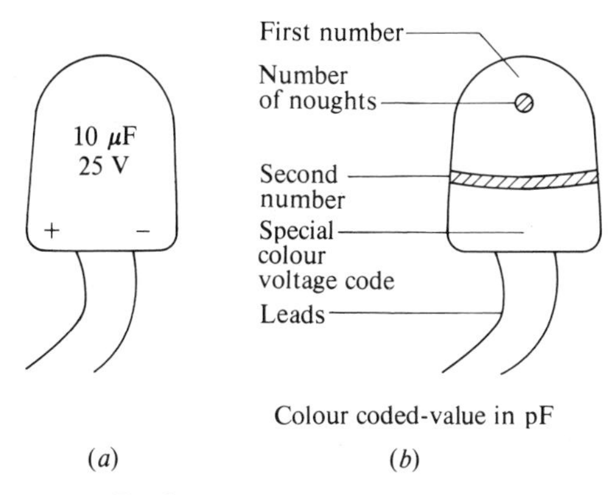

Fig. 3.7 Tantalum capacitors.

3.7 Series and parallel connections

The total capacitance may be modified by connecting capacitors in series and/or parallel.

Note: Take care with voltage ratings.

The total capacitance may be *reduced* by connecting the capacitors in *series*, as shown in Fig. 3.8. The equivalent capacitance C_T may be calculated from:

$$\frac{1}{C_T} = \frac{1}{C_1} + \frac{1}{C_2} + \ldots$$

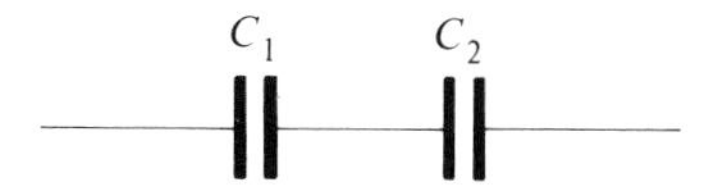

Fig. 3.8 Capacitors connected in series.

The total capacitance may be *increased* by connecting the capacitors in *parallel*, as shown in Fig. 3.9. The equivalent capacitance C_T may be calculated from:

$$C_T = C_1 + C_2 + \ldots$$

3.8 Charging and discharging capacitors

A capacitor may be charged from a d.c. supply through a suitable resistor, as shown in Fig. 3.10. When the switch, S, is closed the voltage V_S causes a current

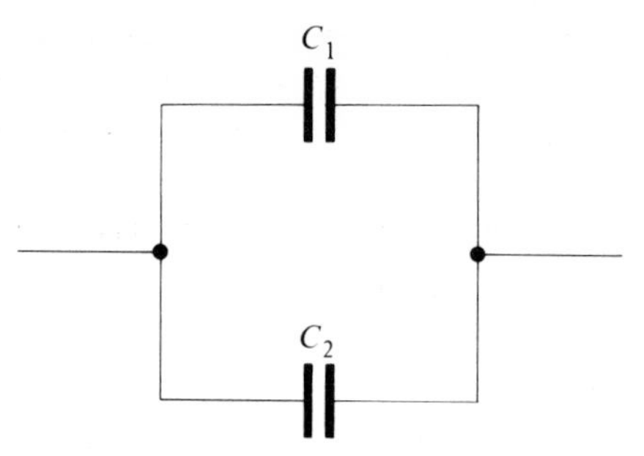

Fig. 3.9 Capacitors connected in parallel.

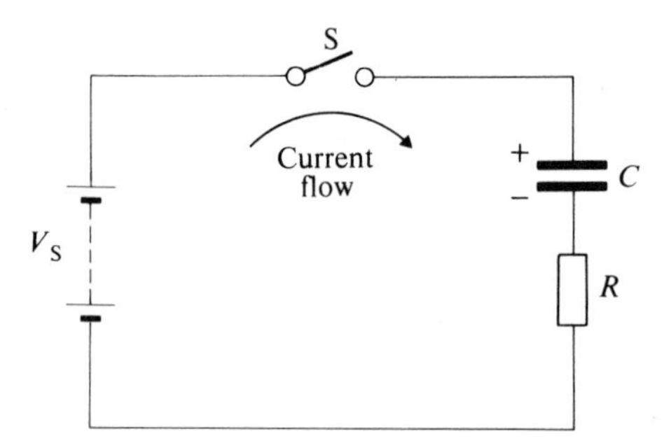

Fig. 3.10 Capacitor charging circuit.

to flow into one side of the capacitor and away from the other. Due to the presence of the insulating dielectric a steady current cannot be set up, so the current decreases as the charge on the capacitor builds up, until $v_C = V_S$ when $i = 0$.

The graph of i and v_C follows an exponential form as shown in Fig. 3.11.

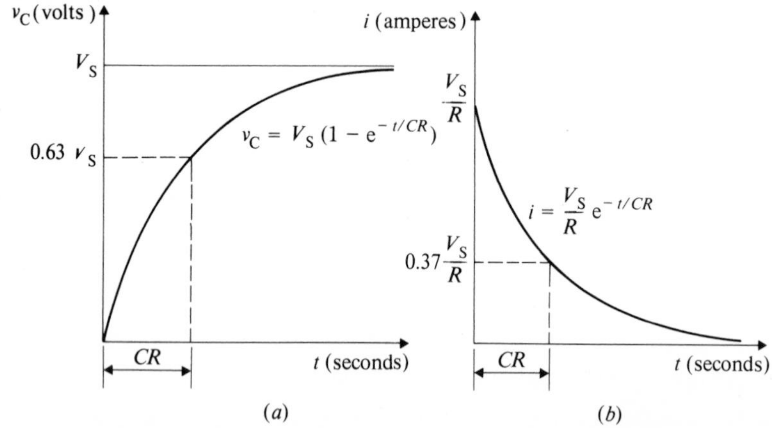

v_C (volts)

V_S

$0.63\,V_S$

$v_C = V_S(1 - e^{-t/CR})$

CR t (seconds)

(*a*)

i (amperes)

$\frac{V_S}{R}$

$i = \frac{V_S}{R} e^{-t/CR}$

$0.37\frac{V_S}{R}$

CR t (seconds)

(*b*)

Fig. 3.11 Graphs of capacitor charging voltage (*a*) and current (*b*).

After a time $T = CR$ seconds the voltage will have risen to 0.63 V_S volts, and the current will have fallen to 0.37 V_S/R amperes. The product CR is called the *time constant* of the circuit, and is widely used in timing and oscillator circuits.

Rule of thumb: After a time of 5 CR seconds, it may be assumed that

$$v_C = V_S \text{ and } i = 0.$$

To show that $CR \equiv T$:

$$C = Q/V, \qquad R = V/I, \qquad \text{and } I = Q/t$$

Therefore,

$$R = Vt/Q.$$

$$CR = \frac{Q}{V} \times \frac{Vt}{Q},$$

and, hence,

$$CR = t$$

where C is in farads and R is in ohms.

Note: C (μF) x R (MΩ) gives T (s).

Discharging the capacitor takes on a similar form. With the supply removed, the capacitor–resistor circuit is connected by a shorting link, as shown in Fig. 3.12. When switch, S, is closed, current flows out of one side of the charged capacitor and back into the other side. When v_C reduces to zero the current ceases. The discharge curves of v_C and i are also of exponential form, as shown in Fig. 3.13.

Assuming that the capacitor was initially charged to V_S, then both v_C and i will fall to 0.37 V_S volts and 0.37 V_S/R amperes, respectively, after a time $T = CR$ seconds.

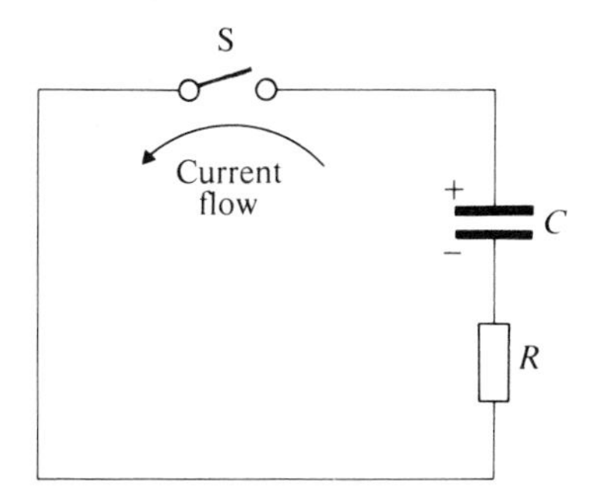

Fig. 3.12 Discharging the capacitor.

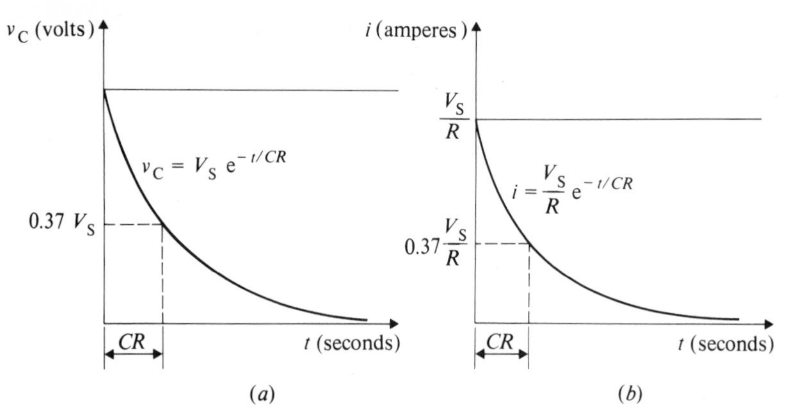

Fig. 3.13 Graphs of capacitor discharging voltage (*a*) and current (*b*).

3.9 Capacitors in a.c. circuits

When connected in d.c. circuits, capacitors charge up to the d.c. voltage and then 'block' any further current flow. In a.c. circuits they are continually charging and discharging and an alternating current does flow in the circuit.

The opposition to the flow of alternating current is called the *capacitive reactance* (symbol X_C) and is measured in ohms:

$$X_C = V_C/I_C\,\Omega$$

and

$$X_C = 10^6/2\pi fC\,\Omega,$$

where C is the capacitance in microfarads (μF) and f is the frequency in hertz (Hz).

Practical Exercise 3a

Simple capacitor test

A quick check to see if a capacitor is functioning is to charge it and discharge it through an ammeter and voltmeter, respectively, or to use a multimeter such as the AVO Multimeter Model 8, as shown in Fig. 3.14.

Note: This is not intended to be conclusive, since it is not possible to test the capacitor to its full rated voltage.

1. Set the multimeter to the Ω x 100 range (i.e., high voltage battery 9 V) and charge the capacitor as shown. *Watch the polarity for electrolytics!*
2. Note the 'kick' on the needle as the charging commences.
3. Disconnect the charged capacitor. Change over the *red* and *black* leads and set the multimeter to the 10 V d.c. range.

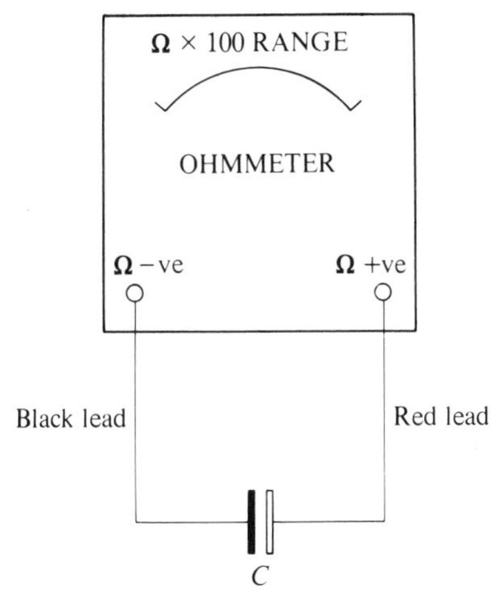

Fig. 3.14 Capacitor testing.

4. Connect the charged capacitor to the multimeter (voltmeter) and watch the voltage indication fall to zero.
5. Repeat for several capacitors.

Did the size of the 'kick' depend on the size of the capacitor? What limited the charge and discharge rates?

4. Inductors and Transformers

4.1 Introduction

An inductor is a coil of wire on a former, and may have a core of air, iron or ferrite. When an electric current flows in the coil, a magnetic flux is produced in the core.

Inductors are widely used in industrial electronics and, as with all electromagnetic devices they are inherently very reliable.

4.2 Self-inductance

When a changing electric current flows in a coil, a changing magnetic field is produced which links with (cuts) the conductors of the coil and causes an e.m.f. to be produced – or induced. This e.m.f. always acts such that it opposes whatever force produced it – the changing current in this case – and is referred to as the *'back' e.m.f.* The magnitude of the back e.m.f. depends on the number of turns on the coil and the rate of change of current.

The *self-inductance* (or simply *inductance*) of a circuit, or of a component is the property in which an e.m.f. is produced when a changing current flows in the circuit, or component.

4.3 Units

The symbol used for inductance is L, and the unit in which inductance is measured is the *henry* (H).

There are two basic methods of defining the 'back' e.m.f. which is induced in a circuit. These are credited to two of our early pioneers, Faraday and Lenz, respectively:

Note: The abbreviation $d\Phi/dt$ means 'rate of change of magnetic flux', i.e., small change of magnetic flux $d\Phi$ in a change in time dt.

$$\text{From Faraday's Law: Induced e.m.f., } e = -N\frac{d\Phi}{dt} \tag{4.1}$$

and,

$$\text{From Lenz's Law: Induced e.m.f., } e = -L\frac{di}{dt} \tag{4.2}$$

Note: the *minus* sign indicates that the induced e.m.f. acts in opposition, i.e., it is a *back e.m.f.*

Now, (4.1) and (4.2) are different ways of expressing the induced e.m.f. in a circuit, and are thus equal to each other:

$$L\frac{di}{dt} = N\frac{d\Phi}{dt}$$

Therefore $$L = N\frac{d\Phi}{di} \tag{4.3}$$

Thus, the self-inductance L is given by:

$$L = \frac{\text{change in flux linkages}}{\text{corresponding change in current}} \tag{4.4}$$

4.4 Stored energy

If the current flowing in an inductor of inductance L rises at a *uniform* rate from zero to I in a time t, then:

Average induced e.m.f., $$e = \frac{LI}{t}$$

and, $$\text{average current} = \frac{I+0}{2} = \frac{I}{2}$$

then, average power = average voltage x average current

Therefore $$P = \frac{LI}{t} \times \frac{I}{2} = \frac{1}{2}\frac{LI^2}{t}$$

The energy used up in producing the magnetic field is therefore *stored* in the magnetic field, and is given by:

Energy = average power x time

$$= \frac{1}{2}\frac{LI^2}{t} \times t = \tfrac{1}{2} L I^2$$

Therefore energy stored, $W = \frac{1}{2} L I^2$ joules (4.5)

If the switch is opened in an inductive circuit, the current tries to fall to zero very quickly and a large e.m.f. is induced by the rapid change of current. Thus, a large amount of energy is stored in the magnetic field, which must be dissipated. This e.m.f. may be large enough to cause the energy to be dissipated in the form of an arc across the switch contacts. In circuits containing high values of inductance, precautions must be taken to prevent damage to switch contacts and other circuit components. A method commonly used is that of switching a resistor into the circuit as the switch is opened, and the energy is dissipated as heat in the resistor.

These induced e.m.f.s exist only when the current is changing, and will be of short duration, i.e., *transient e.m.f.s.*

4.5 Effect of core material on inductance

In addition to the number of turns and the rate of change of flux, the inductance of a coil is dependent upon the dimensions of the coil and the magnetic characteristics of the material used as the core of the inductor.

Ferromagnetic materials are basically iron and iron alloys and have magnetic characteristics which produce a much greater magnetic flux for a given current than an air-cored inductor, as shown in Fig. 4.1.

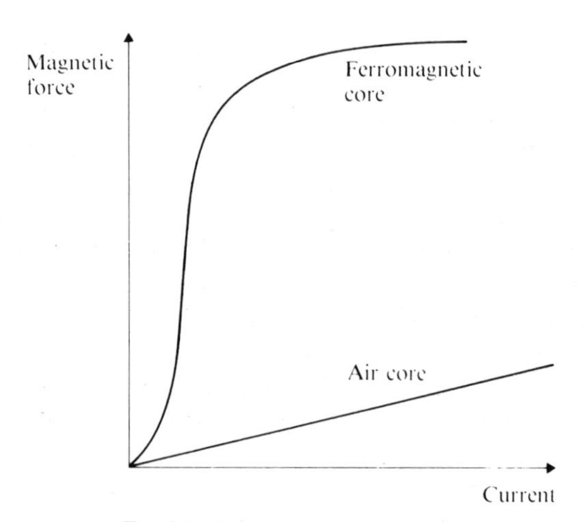

Fig. 4.1 Magnetic characteristics.

The inductance of a given air-cored coil may therefore be considerably increased by using a ferromagnetic core.

A point is reached on the magnetic characteristic when the curve flattens. Any further increase in current is accompanied by only a very small increase in magnetic flux – and the material is *magnetically saturated.*

The ability of a material to *conduct* magnetic flux is termed *permeability*, μ.

$$\mu = \mu_o \mu_r \qquad (4.6)$$

where μ_o = permeability of free space

= a constant

= $4\pi \times 10^{-12}$ H/m

and, μ_r = relative permeability

= a number (depending on point of operation on the magnetic characteristic) – no units.

Note: μ_r for air ≈ 1.

Inductance, $$L = \frac{\mu_o \mu_r A N^2}{l} \tag{4.7}$$

where A = cross-sectional area of coil

N = number of turns on coil

l = length of coil

Since the coil of an inductor is made of wire, it has *resistance* and therefore a limited current-carrying capacity. Also, the inductance of an iron-cored coil is dependent on the current flowing (see the magnetic characteristic in Fig. 4.1). Therefore, inductors are normally quoted as having a given inductance at a given current rating, together with a d.c. resistance.

Ferromagnetic material has an inherently low resistance, so that any e.m.f.s induced in the material will cause localized circulating (*eddy*) currents to flow in the core material. Eddy currents give rise to a power loss which causes heat to be generated in the core. This effect increases with increasing frequency, so that methods must be used to reduce it. *Laminated* (layered) cores are used to increase the core resistance, and thus reduce the effects of eddy currents. This is suitable for power frequencies and the lower audio frequencies. At higher audio and the lower radio frequencies, ferrite (dust) cores are used.

Magnetic fields produced by many pieces of industrial equipment which include inductors, electromagnets, and transformers can cut across other electronic components and circuits and cause undesirable e.m.f.s to be induced. This causes electrical interference – referred to as *electrical noise* – and may prevent the normal operation of the circuit.

It is not possible to obtain a material that can act as an effective magnetic insulator, so that the only available method of containing a magnetic field within a certain area is to provide a high-permeability path for the magnetic flux. This is called a *magnetic shield* (or *screen*).

Magnetic shields are made of high permeability, ferromagnetic material, and can be used to contain the magnetic field produced by, say, a coil, to a given area, or to isolate a component from any stray magnetic field, as shown in Fig. 4.2.

4.6 Growth and decay of current in inductors

In practice, a coil having self-inductance also has some resistance, and this is generally represented in a circuit as two separate components, a coil of self-

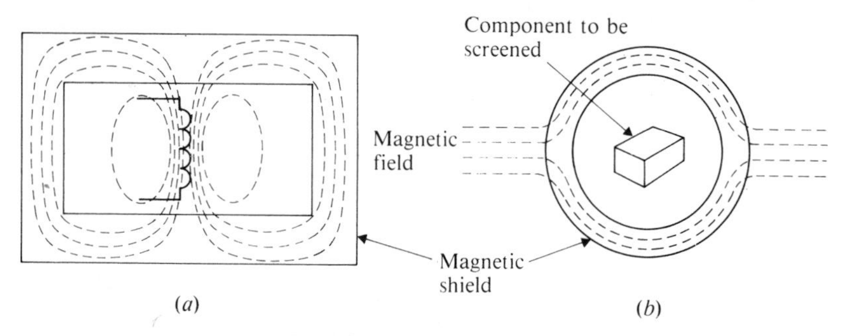

Fig. 4.2 Magnetic shielding.

inductance L, in *series* with a resistance R as shown in Fig. 4.3, where the circuit is connected to a source of d.c. supply, V.

If the self-inductance of the circuit in Fig. 4.3. is negligible, the current will rise instantaneously to a value $I = V/R$. But, when the self-inductance is not negligible, as soon as current begins to flow, a 'back e.m.f.' is induced in the inductance, which retards the growth of current. Thus, the rate of change of current will depend on the total inductance of the circuit.

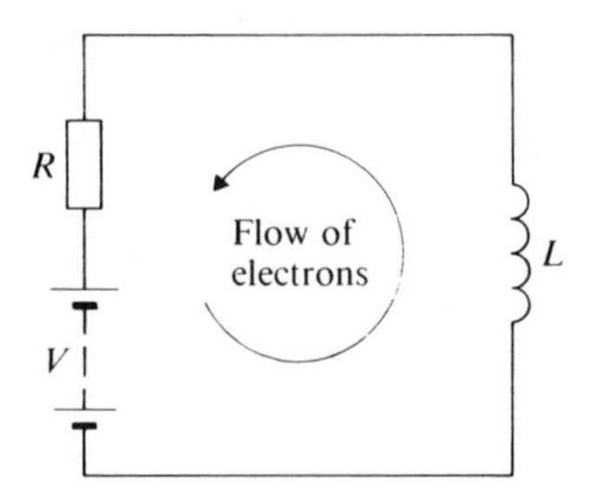

Fig. 4.3 Growth of current in inductor.

Initially, the current is zero, and there is no induced e.m.f. in the inductance.

At switch-on, the current starts to rise. This change of current induces an e.m.f. in the inductance, which opposes the applied voltage. Thus, the initial rate of change of current is 'slowed up'. The current continues to rise, but the rate of change is gradually retarded, so that the induced e.m.f. becomes smaller until, finally, the current reaches its maximum value of V/R and the induced e.m.f. will be zero, as shown in Fig. 4.4.

Let the current at any instant $= i$

and, the induced e.m.f. at that instant $= e$

then $V = e + iR$

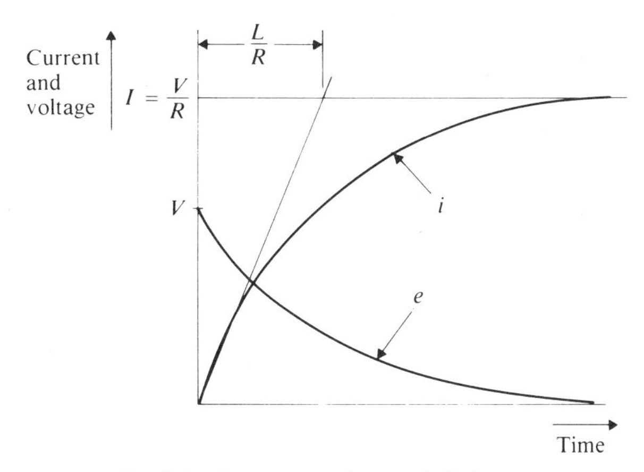

Fig. 4.4 Current growth curve in inductor.

As the current i rises, the induced e.m.f. e must become smaller, but since e depends upon the *rate of change of current*, then the rate of change of current becomes slower – until theoretically after infinite time i reaches I (and e reaches zero) but, in practice, i and e are near enough to these conditions within seconds, or minutes, depending on the values of L and R.

The time constant is the time that would be taken for the current to reach its maximum value if the initial rate of change of current were maintained.

Alternatively, it is the time taken for the current to reach 63.2 per cent of its final steady value – when the rise is exponential, as in practice.

Time constant, $T = \frac{L}{R}$ seconds

where L = self-inductance (henrys)

R = total circuit resistance (ohms)

If the current flowing in the circuit shown in Fig. 4.3 is allowed to reach its steady value of I, the magnetic field around the coil will no longer be changing. If the source of supply is now disconnected and simultaneously replaced by a short circuit as shown in Fig. 4.5, the tendency will be for the current to fall to zero, but this cannot take place instantaneously, because as soon as the value of current changes, an opposition e.m.f. is induced in the inductor, which tends to maintain the flow of current in the circuit. The decay of current is thus delayed, and the induced e.m.f. will decrease from V volts to zero, as shown in Fig. 4.6.

The time constant must now be defined as the time that would be taken for the current to reach zero, if the initial rate of change of current were maintained.

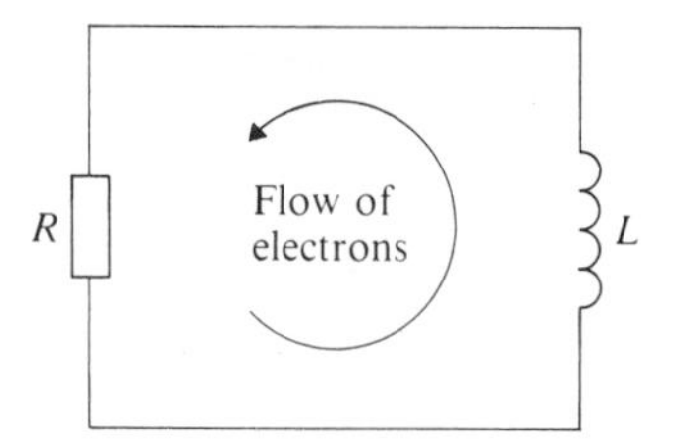

Fig. 4.5 Decay of current in inductor.

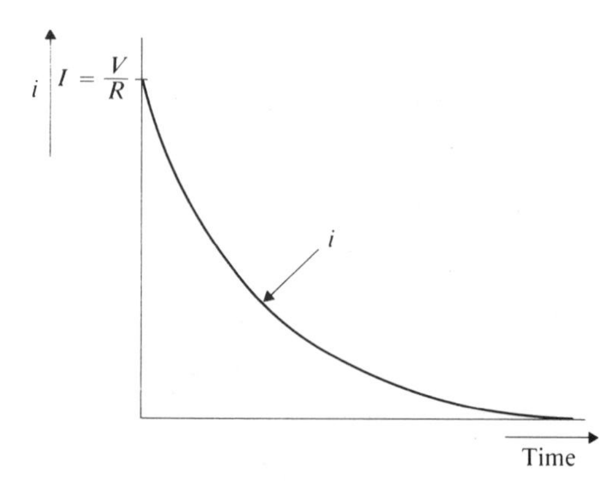

Fig. 4.6 Current decay in inductor.

Alternatively, it is the time taken for the current to decay to 36.8 per cent of its original value – when the decay is exponential, as in practice.

Time constant, $T = \frac{L}{R}$ seconds

Practical Exercise 4a

Induced e.m.f. in inductor

Connect up the circuit shown in Fig. 4.7, in which the coil (inductor) may be a secondary winding from the a.c. power supply circuit on the Electronic Learning Kit.

Note: Ensure that the a.c. power supply switch is in the OFF position throughout this test.
Adjust the d.c. power supply to approximately 1.5 V.
Switch ON the d.c. power supply to the circuit.
Observe the Neon as the switch is opened.

The Neon will glow momentarily – indicating that the voltage across it must

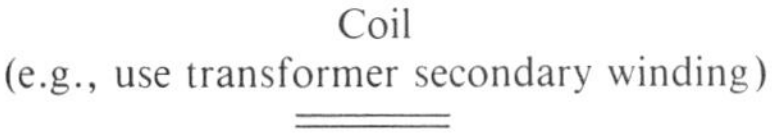

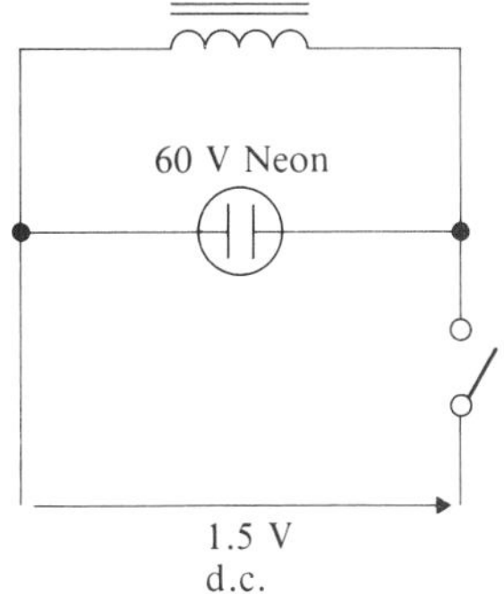

Fig. 4.7 Induced e.m.f. in inductor.

be at least 60 V to strike the neon – even though the d.c. voltage originally applied to the circuit was only 1.5 V.

4.7 Mutual inductance

Two circuits are said to possess *mutual* inductance if a *changing current in one circuit* gives rise to a changing magnetic flux, which links with the second circuit, causing an *e.m.f. to be induced in the second circuit.*

The mutual inductance of two circuits is *1 henry* if the current in one circuit changing at the rate of 1 ampere/second causes an e.m.f. of 1 volt to be induced in the second circuit. Thus

$$e = -M\frac{di}{dt} \tag{4.8}$$

where M = mutual inductance (henrys)

$\frac{di}{dt}$ = rate of change of current in first circuit.

the minus sign indicates that the induced e.m.f. acts in such a direction as to cause current to flow in the second circuit which will oppose the flux set up by the first circuit.

Therefore
$$e = -N_2\frac{d\Phi}{dt} = -M\frac{di}{dt}$$

So
$$M = N_2\frac{d\Phi}{di} \tag{4.9}$$

or
$$\text{Mutual inductance} = \frac{\text{change in flux linkages in second circuit}}{\text{change in current in first circuit}} \tag{4.10}$$

4.8 Inductance of coils connected in series

When two coils are connected in series in a circuit, and there is a mutual inductance between the coils, then the total equivalent inductance depends on

1. The self-inductance of each coil.
2. The mutual inductance between the coils.
3. The method of winding of the coils relative to each other.

Two coils of self-inductance L_1 and L_2, connected in series, have a mutual inductance M between them, as shown in Fig. 4.8.

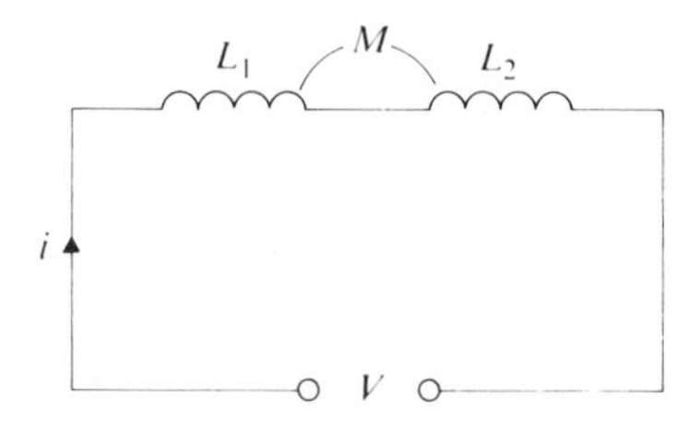

Fig. 4.8 Mutual inductance.

When the current through the circuit changes, the effective induced e.m.f. in L_1 will be the algebraic sum of the self-induced e.m.f. in L_1 and the mutually induced e.m.f. from L_2. The effective induced e.m.f. in L_2 can similarly be found. If the magnetic flux set up by changing current in L_1 *assists* the flux set up in L_2, then the coils are connected in *series aiding*, and the total effective inductance is given by:

$$L_T = L_1 + L_2 + 2M \tag{4.11}$$

However, if the two coils are wound so that for a given current flowing in the series circuit their magnetic fluxes *oppose* one another, then the coils are connected in *series opposition*, and the total effective inductance is given by:

$$L_T = L_1 + L_2 - 2M \tag{4.12}$$

4.9 The transformer

Transformers are basically made up from two coils, electrically insulated from one another, which are wound over a common core material formed into a closed magnetic circuit, so that the two coils are mutually coupled.

When the transformer is used at power frequencies and frequencies up to the audio limit, the magnetic core is made up of laminations of high permeability material. When used at radio frequencies, the losses due to eddy currents become excessive if laminations are used, and ferrite or dust cores may then be used.

The *primary* coil is connected to an a.c. source and the *secondary* coil is provided with terminals from which the output alternating voltage is taken.

Transformers are used in a number of applications:

1. To change the level of alternating voltage, either *step-up*, or *step-down*.
2. Impedance matching.
3. Coupling.
4. Insulation between circuits.

Transformation Ratio, $n = \dfrac{V_p}{V_s} = \dfrac{N_p}{N_s} = \dfrac{I_s}{I_p}$

Consider the resistive-loaded transformer shown in Fig. 4.9.

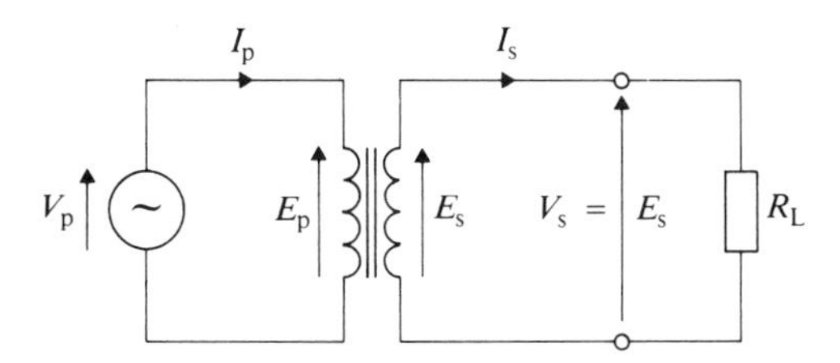

Fig. 4.9 Transformer on resistive load.

If the resistance presented by the primary circuit = R_1

then
$$R_1 = \frac{V_p}{I_p}$$

Similarly, the total secondary resistance R_L is given by

$$R_L = \frac{V_s}{I_s}$$

Therefore
$$\frac{R_1}{R_L} = \frac{(V_p/I_p)}{(V_s/I_s)} = \frac{V_p}{V_s} \cdot \frac{I_s}{I_p} = n^2$$

Thus equivalent input resistance, $R_1 = n^2 R_L$

4.10 Constructional Features

1. ***Magnetic circuit.*** Three types of construction are widely used. In the *shell* type the laminations are arranged as shown in Fig. 4.10(a), where both primary and secondary are wound on a common central limb, with the outer limbs completing the high permeability path. The *core* type has the laminations arranged as shown in Fig. 4.10(b), where part of each winding

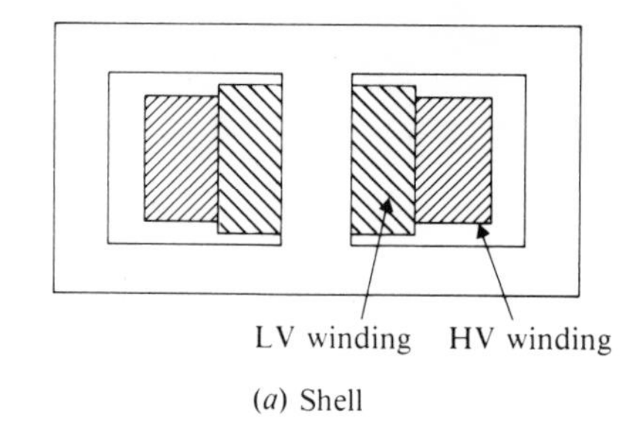

(*a*) Shell

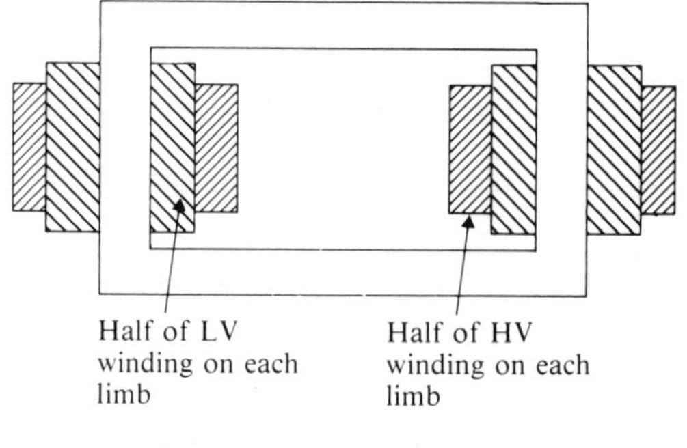

(*b*) Core

Fig. 4.10 Magnetic circuit arrangements of transformers.

is wound on each limb to reduce the effect of flux leakage. Note that owing to the problem of insulation, the low-voltage winding is usually wound so that it is nearest to the core. Finally, the *toroidal* transformer features the ideal shape to use the minimum amount of material. Small size and weight (approximately 50 per cent of conventional transformers), low noise, and low magnetic interference field make the toroidal transformer ideal for compact power supplies. A higher flux density is possible since the magnetic flux is in the same direction as the rolling direction of the grain-orientated core plate. Iron losses are very small, and with no air gaps and no loose sheets which can vibrate, the 'mains hum' commonly found in transformers is virtually eliminated. All windings are symmetrically spread over the entire core, and a high current density is possible since the whole surface of the toroidal core allows efficient cooling of the windings.

2. ***Windings.*** Two forms of winding are widely used. In *concentric coil* the windings are wound as a complete coil – the high-voltage winding being wound around the outside of the lower-voltage winding, as shown in Fig.

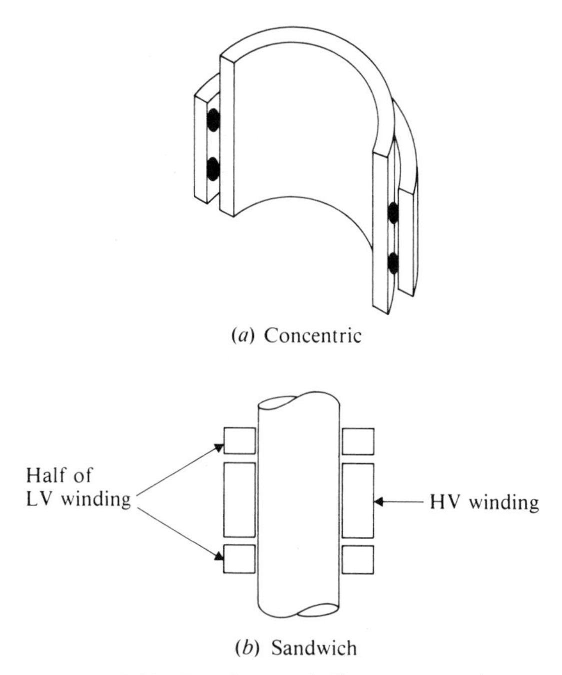

Fig. 4.11 Transformer winding arrangements.

4.11(a). The *sandwich* winding tends to reduce leakage flux, and the windings are arranged in the form of a 'sandwich' as shown in Fig. 4.11(b).

3. ***Cooling.*** Two methods of cooling are used depending on the size and application of the transformer. *Air cooling* is generally used with small, low-power transformers. High-voltage and high-power transformers are generally enclosed in steel tanks, and *oil* is used as the cooling medium.

5. Rectification, Power Supplies, and Smoothing

5.1 Introduction

Modern diodes are made from semiconductor material. Early versions were derived from germanium because of the ease with which the basic material was purified compared with silicon. However, all germanium devices suffer from the same major disadvantages, i.e., they are all destroyed by very moderate temperature rises. As soon as purification of silicon to the required degree was practicable, silicon devices began to take over; now they completely dominate the discrete semiconductor market.

5.2 Diodes

These are two-terminal devices formed by creating a junction between two types of semiconductor, p type and n type silicon. The resulting device will pass current relatively easily in one direction but hardly at all in the reverse direction.

Note: the arrowhead on the symbol shown in Fig. 5.1 points in the direction of easy current flow – this is a general rule for the circuit symbols of all semiconductor devices.

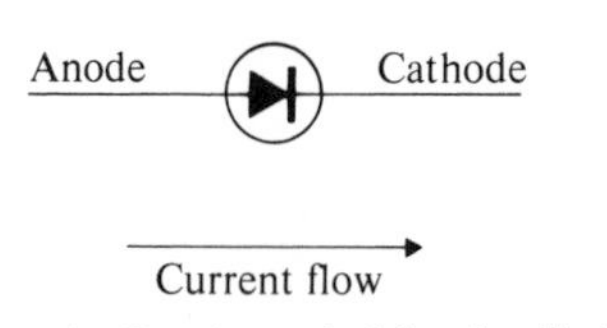

Fig. 5.1 Circuit symbol for the diode.

Diodes are manufactured in a variety of shapes and sizes and are put to many uses. Some general shapes are illustrated in Fig. 5.2. Note the band at the cathode end in the first, and the bullet-shaped one which points in the direction of the

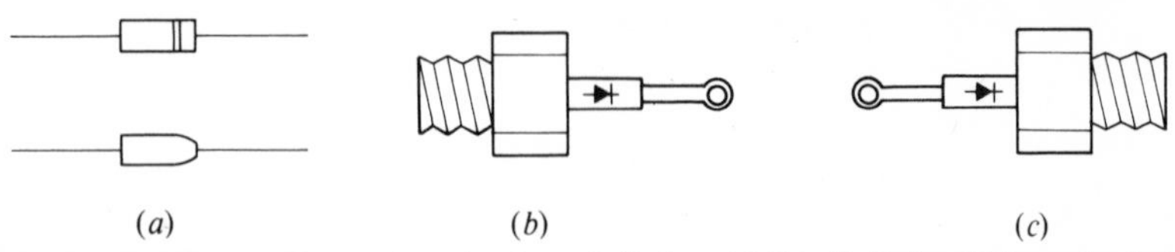

Fig. 5.2 General outlines of a selection of diodes. (*a*) Small diodes; (*b*) anode stud; (*c*) cathode stud.

current flow. Larger diodes, which are capable of dissipating more power, may be made with a stud as one of the terminals. This allows connections direct to the heat sink. Now because of the different ways that we connect diodes, it is sometimes desirable to connect the cathode of one and the anode of another to the same heat sink – hence, the cathode and anode stud types. Watch it! If the symbol is not visible, test the diode to make sure which type it is before connecting a replacement diode in circuit.

Practical Exercise 5a

Testing a diode

To test a diode for direction of current flow and, hence, identify its terminals, proceed as follows:

1. Connect the 12 V d.c. supply to a 12 V lamp, switch on, and note the brilliance of the lamp.
2. Add the diode in series with the lamp, as shown in Fig. 5.3, first one way round and then in reverse. In each case, measure the voltage across the lamp and the voltage across the diode.

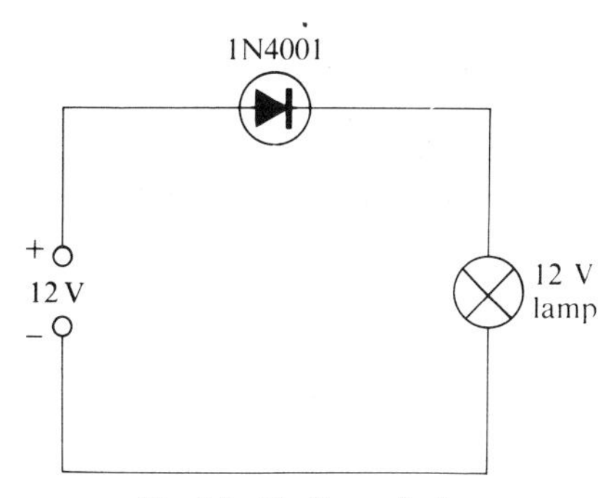

Fig. 5.3 Testing a diode.

   (a) ***When the lamp lit:***
   The voltage across the lamp was ________V.
   The voltage across the diode was ________V.
   Was the lamp as bright as before? Why?

   (b) ***When the lamp did not light:***
   The voltage across the lamp was ________V.
   The voltage across the diode was ________V.

3. Using the analogue type of multimeter, such as the AVO Multimeter, Model 8, select the $\Omega \div 100$ range (i.e., the low-voltage battery, 1.5 V).
   Connect the diode to the ohmmeter terminals as shown in Fig. 5.4.(*a*),

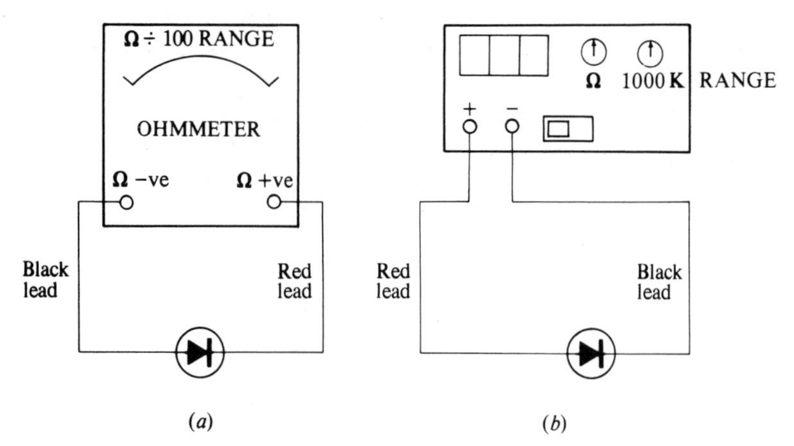

Fig. 5.4 Testing a diode with the ohmmeter. (*a*) Using analogue ohmmeter; (*b*) using DMM.

then reverse it. Current flows when the ohmmeter positive terminal is connected to the ________ and the ohmmeter negative terminal to the ________.

4. Using the digital multimeter (DMM), select the Ω x 1000K range (i.e., low voltage output). There is no need to reverse the leads when using the DMM, the polarities of the terminals are correctly identified.

   Connect the diode to the DMM as shown in Fig. 5.4(*b*), then reverse it. Note the results.

Note: It is suggested that when using ohmmeters for testing semiconductor devices in this way that the *low voltage range is always used*, since some devices are very vulnerable to voltages above 5 V (see Chap. 6).

5.3 Diode ratings

Several ratings are quoted for the diode; these include:

1. ***Repetitive reverse maximum voltage,*** V_{RRM}. When a diode is connected so as to block current flow, i.e., when it is *reverse biased*, only a small *leakage* current should flow (a few nanoamperes for silicon, and up to about 50 μA for germanium). During this time the reverse voltage causes electrical stress to be applied to the diode. If this is too great for the diode it will break down, conduct, and be destroyed. The peak inverse voltage that a diode can stand is V_{RRM}.
2. ***The maximum steady current that a diode can pass in the forward direction,*** $I_{F(max)}$. The power dissipated by the diode in this manner is $V_F I_F$. Now, for silicon devices, $V_F \approx 0.7$ V and for germanium $V_F \approx 0.3$ V, so that $V_F I_{F(max)}$ is an indication of the maximum power dissipation that the diode can handle. This can be extended by artificial cooling, e.g., heat sink,

air blast, water cooling through pipes surrounding the diode. Many silicon diodes will work with junction temperatures well over 200 °C, i.e., outside case temperature approaching 150 °C.

Some diodes, particularly those used in 'smoothed' power supplies, are only required to pass current for short periods during each cycle. These currents are generally several times higher than the steady load current and there is thus a need for a surge rating for diodes.

5.4 Power supplies

Several simple arrangements of diodes yield a number of d.c. power supplies from a.c. sources.

1. ***Half wave.*** As shown in Fig. 5.5. This arrangement is not much used.

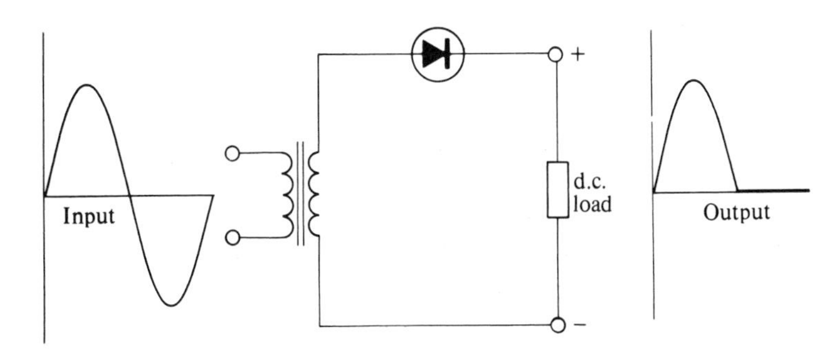

Fig. 5.5 Half-wave rectified a.c. power supply.

2. ***Full wave.*** As shown in Fig. 5.6. Uses both half-cycles of the a.c. supply, but needs a centre-tapped transformer rated at $2V_L$. The diodes must also have the same V_{RRM} rating.

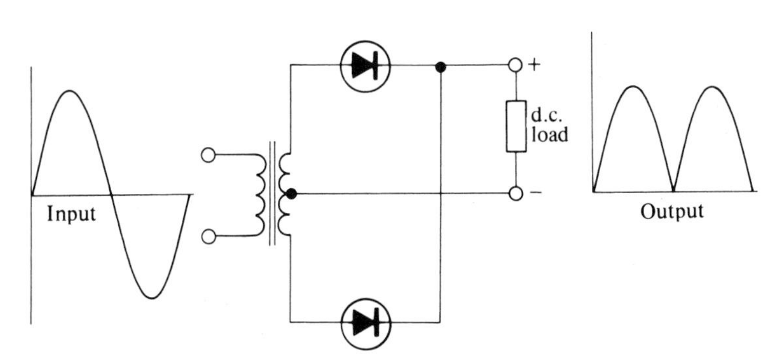

Fig. 5.6 Full-wave rectified a.c. power supply.

3. ***Full-wave bridge.*** As shown in Fig. 5.7. Uses four diodes, but is generally cheaper because of the simpler transformer rated at V_L. The diodes conduct in parallel pairs on alternate half-cycles. The diode V_{RRM} rating is $-V_L$.

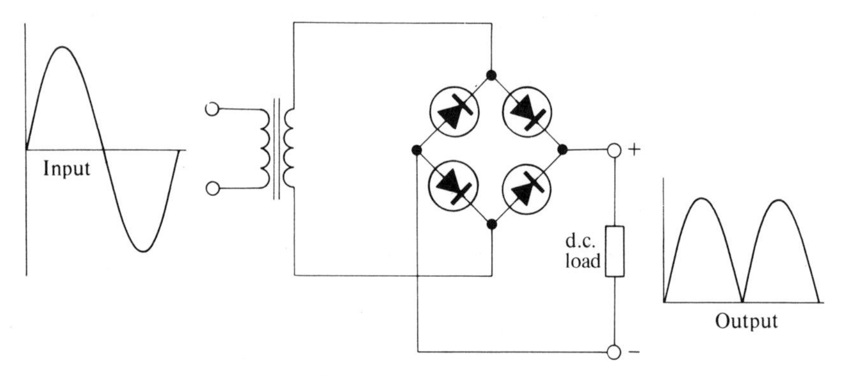

Fig. 5.7 Full-wave bridge rectifier circuit.

5.5 Smoothing

In order to *smooth* the rectified a.c. supply to a small 'ripple', a smoothing filter is used. This can consist of a single large electrolytic capacitor connected across the load terminals as shown in Fig. 5.8 or may consist of a π type filter as shown

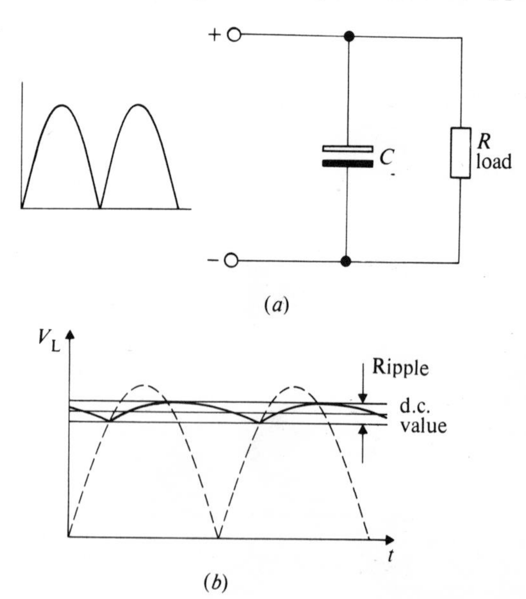

Fig. 5.8 Reservoir capacitor for smoothing. (*a*) Circuit; (*b*) waveforms.

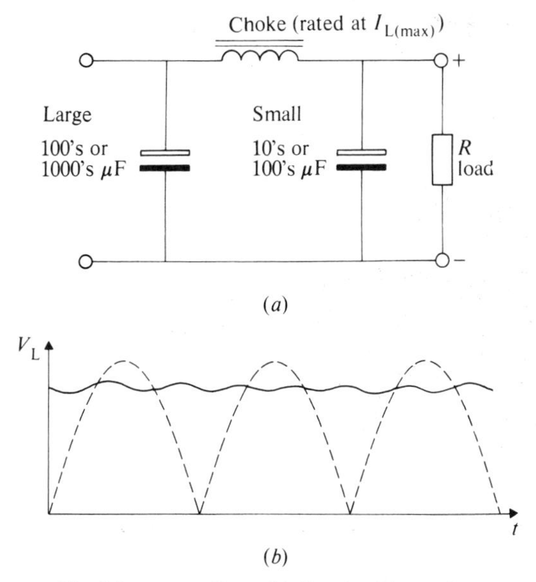

Fig. 5.9 π type filter. (a) Circuit; (b) waveform.

in Fig. 5.9. The object of *smoothing* is to raise the average level of d.c. voltage and to keep the percentage ripple below the maximum specified at full load.

Note: the ripple frequency will be twice the supply frequency for full-wave rectification.

Practical Exercise 5b

Power supplies and smoothing

Connect up the circuit as shown in Fig. 5.10, with a 1K0 load resistor:

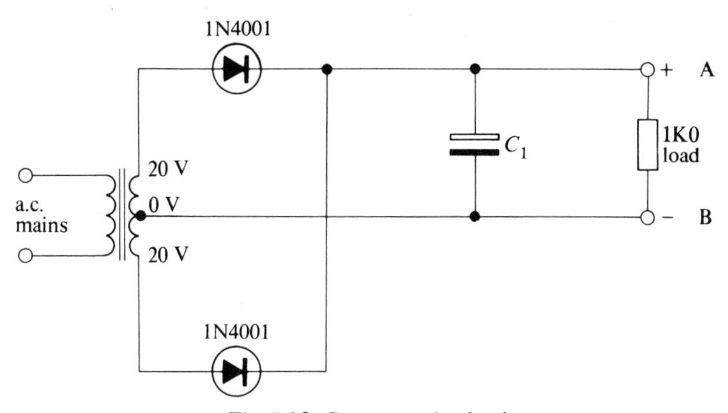

Fig. 5.10 Power supply circuit.

1. Measure the voltage across AB using the CRO and examine the waveform:
   (a) with C_1 excluded from the circuit.
   (b) with C_1 included in the circuit.
2. Repeat using different values of capacitor. Which value of capacitor gives the smallest a.c. ripple across the load resistor?
3. Repeat the above measurements with the series 'choke' and second capacitor connected to form the π filter.
   (a) The peak a.c. voltage applied to the circuit = ________ V.
   (b) The steady d.c. voltage across the load = ________V.
   (c) The minimum a.c. ripple voltage across the load = ________V peak to peak.
   (d) The minimum a.c. ripple voltage was obtained with C_1 = ________ μF, C_2 = ________ μF, and L_1 = ________ H.

Practical Exercise 5c

Simple transistor radio d.c. power supply

Connect up the simple power supply circuit shown in Fig. 5.11, where the load is a transistor radio requiring 9 V d.c.

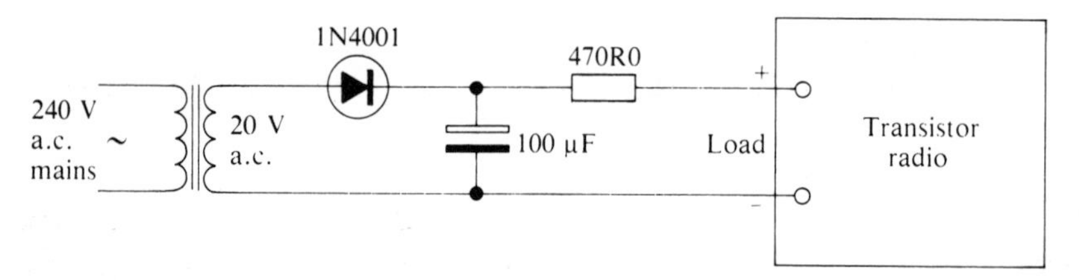

Fig. 5.11 Transistor radio simple d.c. power supply.

Switch on the power supply and tune in the radio. Note the distortion that occurs in the sound. This is mainly due to the unsteady nature of the power supply voltage.

Practical Exercise 5d

Regulation of transformer a.c. power supply

Connect up the circuit shown in Fig. 5.12.

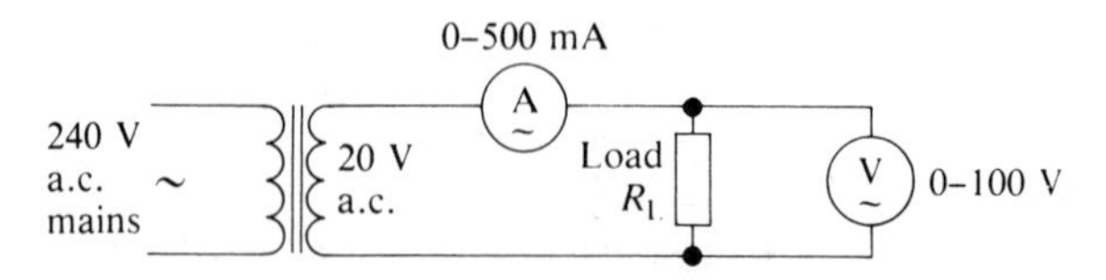

Fig. 5.12 Transformer voltage regulation test.

Record values of load resistance, load voltage, and load current as the resistance is changed from 10K0 down to 100R0.

Plot a graph of current/voltage over the complete range.

Comment on the shape of the graph.

Practical Exercise 5e

Regulation of simple d.c. power supply

Connect up the circuit shown in Fig. 5.13.

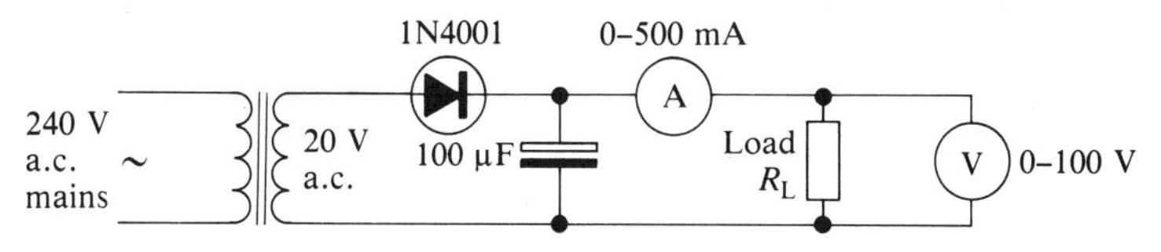

Fig. 5.13 Voltage regulation of simple d.c. power supply.

Record values of load resistance, load voltage, and load current as the resistance is changed from 10K0 to 100R0.

Plot a graph of current/voltage over the complete range.

Compare the shape of the graph with that obtained from Practical Exercise 5d.

5.6 Voltage doublers

Several requirements occur in electronics for voltages which are greater than that normally available. The principle of a *voltage doubler* circuit is shown in Fig. 5.14(*a*). During the first half cycle of the a.c. supply diode D_1 charges capacitor C_1 to approximately V_{max}. During the second half cycle, diode D_2 charges capacitor C_2 to V_{max}. Since C_1 and C_2 are connected in series, V_{out} is approximately *twice* V_{max}.

Alternatively, in Fig. 5.14(*b*), C_1 is charged through D_1 during the negative half cycle of the supply. This d.c. potential is now in series with the a.c. supply so that during the positive half cycle C_2 is charged to a voltage equal to the sum of V_{max} and the voltage across C_1. Thus, V_{out} is approximately equal to *twice* V_{max}.

However, when loaded, each of these circuits produces a large ripple voltage. This means that their use is restricted to low-current applications.

Practical Exercise 5f

Voltage doublers

Connect up the circuit as shown in Fig. 5.14(*a*), with a 10K0 load resistor:

1. Measure V_{out} using a voltmeter *and* the CRO and examine the waveform.

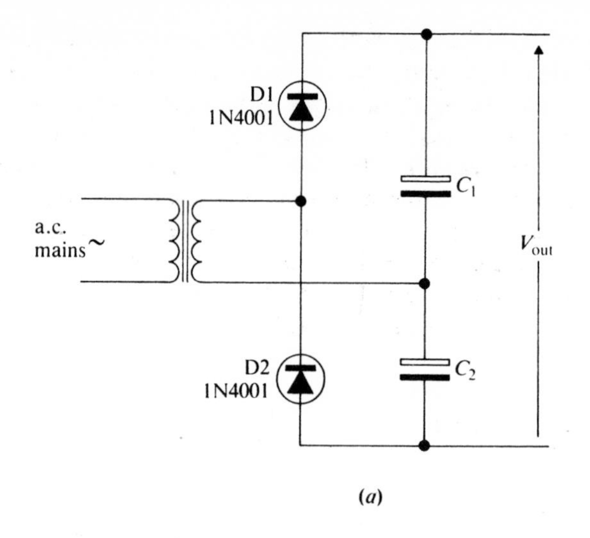

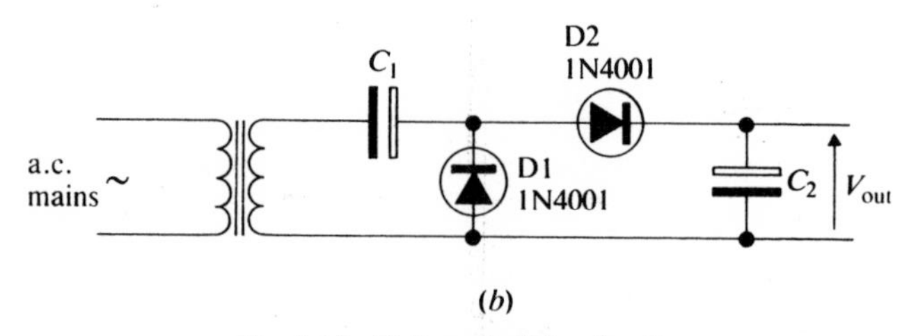

Fig. 5.14 Voltage doubler circuits.

Record the voltage and the waveform.

V_{out} = ________ volts

2. Measure the voltage across C_1 and across C_2.
   Note: Take care with the polarity of the voltmeter!

   V_{C1} = ________ volts

   V_{C2} = ________ volts.

3. Compare ($V_{C1} + V_{C2}$) with V_{out}.
4. Repeat (1) to (3) with a 1K0 load resistor.
5. Comment on the effect of the increased load.
6. Repeat (1) to (5) with the circuit connected as shown in Fig. 5.14(*b*).

5.7 Zener diodes

Zener diodes are named after the effect discovered by a researcher of the same name. Although the effect is limited to diodes up to only about 6 V, other diodes, having roughly the same characteristics, but working on different principles, are commonly called *Zener diodes*. The circuit symbol of the Zener diode is shown in Fig. 5.15(*a*) and the characteristic is shown in Fig. 5.15(*b*).

These diodes have a normal forward characteristic, i.e. they pass current as normal diodes when forward biased. When reverse biased they block in the same way, until the *Zener* voltage is reached when a sudden breakdown occurs. The difference between this breakdown and the destructive breakdown which occurs in a normal diode, is that if the reverse current is limited by external resistance to less than a rated value, then the diode will survive.

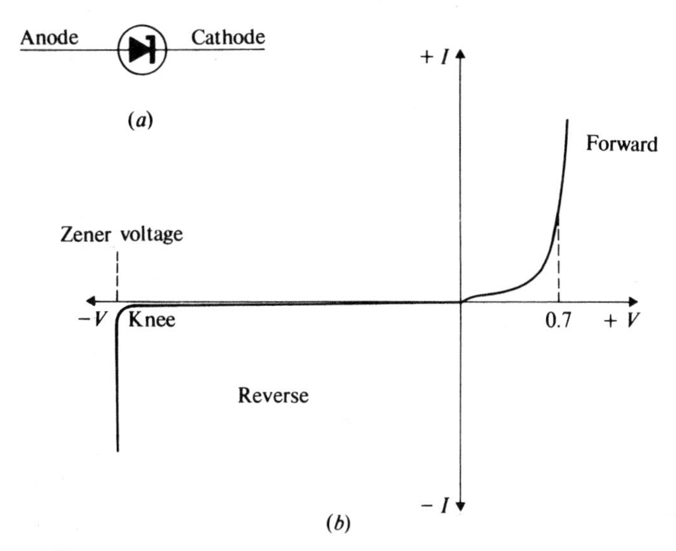

Fig. 5.15 The Zener diode. (*a*) Circuit symbol; (*b*) characteristic.

In a good Zener diode the characteristic will be almost right-angled, i.e., the Zener voltage will be constant. This leads to their use as *reference* diodes.

5.8 Zener diode ratings

1. ***Zener voltage.*** Zener diodes are made in a range of preferred values of voltage:

   3.3, 4.7, 5.1, 6.2, 6.8, 9.1, 10, 11, 12, 13, 15 up to 200 V.

2. ***Power rating.*** More power will be dissipated in the reverse direction for a given current.

EXAMPLE

A forward current of 1 A gives:

$$P \approx 1 \times 0.7 = 0.7 \text{ W}.$$

A reverse current of 1 A for, say, a 10 V Zener diode, gives:

$$P = 1 \times 10 = 10 \text{ W}.$$

Practical Exercise 5g

Zener diode characteristic

Connect up the circuit arrangement shown in Fig. 5.16.

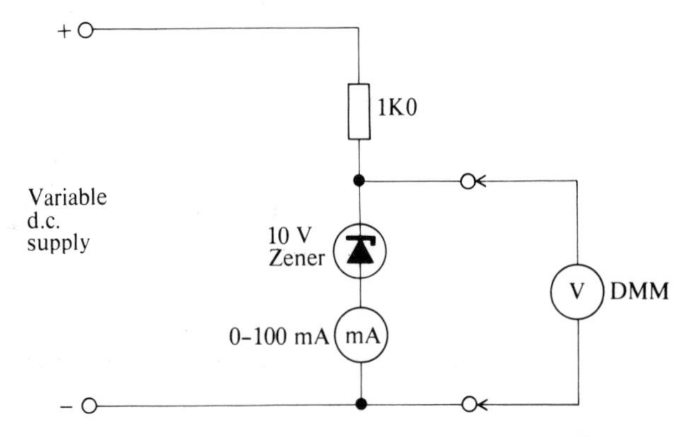

Fig. 5.16 Circuit to obtain Zener diode characteristics.

1. Apply a *forward bias* voltage across the Zener diode, and record values of current and voltage as the supply voltage is varied.

   Note: Do not exceed the Zener power rating.

   Forward characteristic:

   V (V)________.

   I (mA)________.

2. Reverse the Zener diode, apply a *reverse bias* voltage across the Zener diode, and record values of current and voltage as the supply voltage is varied.

   Note: Do not exceed the Zener power rating.

   Reverse characteristic:

   V (V)________.

   I (mA)________.

3. Plot a graph of the above results and compare the shape of the graph to that shown in Fig. 5.15(*a*).

5.9 Reference supply voltage

A reference supply voltage can be obtained to supply a fixed voltage to a load when the normal d.c. supply voltage is subject to variations, e.g., a car electrical system. A typical application is to supply, say, a 6 V transistor radio or tape recorder from the 12 V d.c. car electrical system, as shown in Fig. 5.17. The series resistor R drops the difference in voltage between the supply voltage and the reference (Zener) voltage.

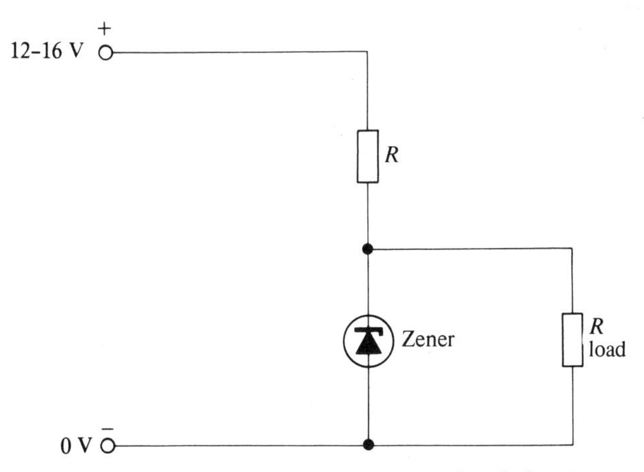

Fig. 5.17 Reference supply voltage – car electrical system.

To find a suitable value of R and its power rating, we choose the *highest* level that the supply voltage is likely to reach, say, 16 V for a nominal 12 V car electrical system, and we choose the smallest possible value of load current, because as the load current decreases, the Zener current increases.

Now, assume that we are going to use a Zener diode, 6V2 rated at 5 W:

$$\textit{Maximum}\text{ Zener current} = \frac{P_Z}{V_Z} = \frac{5}{6.2} = 0.81 \text{ A}.$$

Assume that the *minimum* load current = 0.2 A, i.e., 200 mA in this case.

Note: This minimum load current will, of course, depend on the load that is to be supplied. Refer to manufacturer's data when in doubt.

In this case, the *total* current, $I_T = 0.81 + 0.2 = 1.01$ A. Therefore,

$$R = \frac{V_{max} - V_Z}{I_T} = \frac{16 - 6.2}{1.01} = 9.9\ \Omega.$$

Therefore, a 10 Ω ± 5 per cent would do.

Power rating of resistor $= I^2R = 1.01^2 \times 10 \approx 10$ W.

In this example, the series-limiting resistor would be 10 Ω ± 5 per cent rated at 10 W, and the Zener diode would be 6V2, 5 W.

Note: In the simple circuit shown, if the load was disconnected from the circuit, the Zener current would increase to a disastrous level. Therefore, precautions would be necessary in the method used to connect the load to the circuit and/or switching the power supply to the circuit.

5.10 Stabilized power supplies

The power supplies needed for many electronic devices must be stabilized, i.e., they are required not to change much in value over the whole load range.

EXAMPLE

1. TTL integrated circuits require 5 V ± 0.25 V.
2. Many integrated circuit operational amplifiers require + 15 V, 0, – 15 V ± 1 V.

A simple *Shunt* (parallel) *stabilizing circuit* has already been dealt with above (Fig. 5.17). An improvement can be made to the simple *shunt regulator* by using a power transistor to conduct the large currents as shown in Fig. 5.18(*a*). Any tendency for the output voltage to rise is 'felt' by the Zener diode which causes transistor TR_1 to conduct more and bypass the excess current from the load – thus reducing load current and output voltage. This arrangment therefore reduces ripple voltage (which may be considered as an output voltage variation to be regulated). Additionally, this circuit will correct for variations of voltage due to temperature changes.

A simple series regulator is shown in Fig. 5.18(*b*) in which an increase in load current causes the transistor current to increase. But, since the supply current is reasonably constant, the current through R_1 must fall, which in turn causes the transistor base current to fall which causes the transistor to conduct less, thus reducing the load current. This circuit regulates against changes due to temperature, and also reduces ripple voltage. The output voltage is formed from the sum of the Zener voltage and V_{BE} of the transistor – which are in opposition. Therefore

$$V_{out} = V_Z + (-V_{BE})$$

Thus, an increase in output voltage results in V_{BE} bcoming less negative, i.e., reducing TR_1 conduction, which reduces load current and the output voltage falls.

Variation of output voltage can be achieved by using the arrangement shown in Fig. 5.19, in which transistor TR_2 directly controls the bias of TR_1, which adjusts the output voltage. This circuit is a variation of the series regulator shown in Fig. 5.18(*b*).

Further improvements can be made to the regulating properties by using a

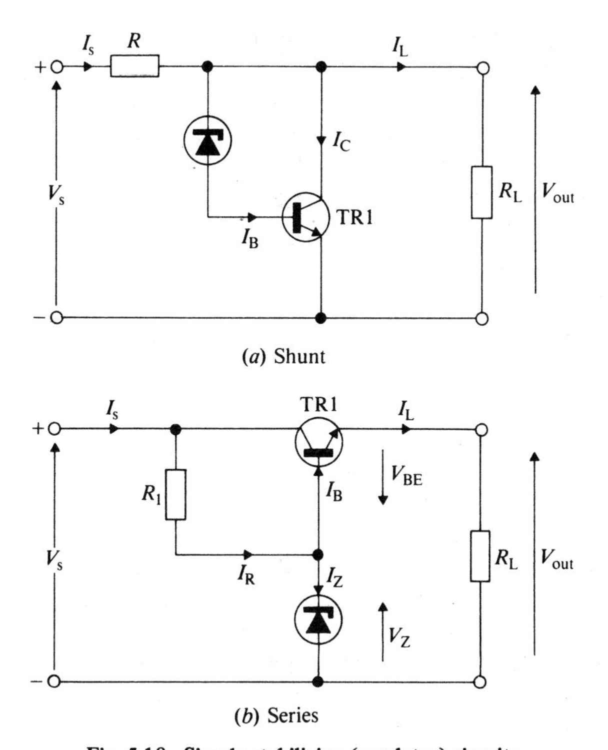

Fig. 5.18 Simple stabilizing (regulator) circuits.

high-gain amplifier (e.g., operational amplifier) as a comparator, as shown in Fig. 5.20. A reference voltage is applied to the non-inverting input of the high gain amplifier and part of the output voltage is fed back to the inverting input of the amplifier. Any difference between these two voltages is amplified (by several thousand times) to create the control voltage for transistor TR_1.

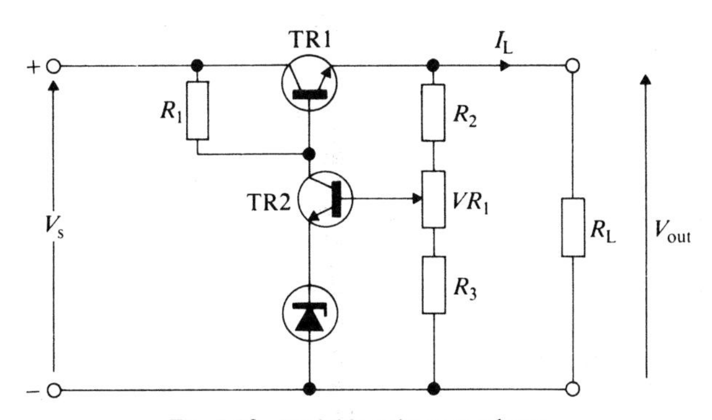

Fig. 5.19 Variable-voltage regulator.

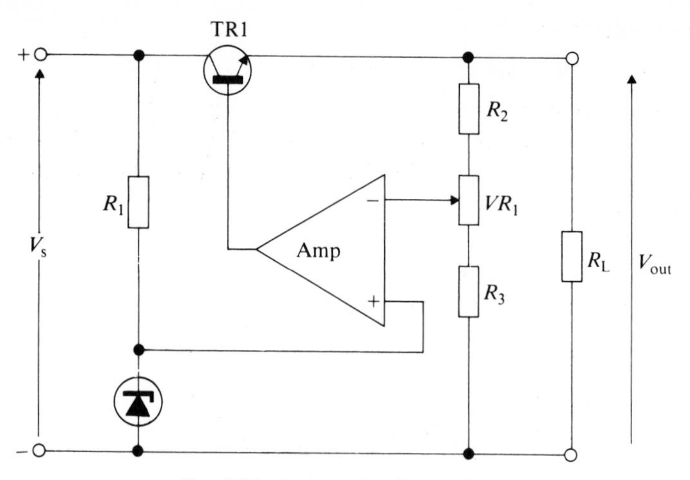

Fig. 5.20 Improved series regulator.

These circuits may often take the form of a single integrated circuit power regulator enclosed in one three-terminal package. In addition to maintaining a constant voltage over the load range the regulator may contain a *crowbar* protection device. The output voltage falls only slightly on load until the maximum load value is reached and then the regulated voltage rapidly collapses to a very small value, i.e., automatic short-circuit (crowbar) protection is provided as shown in Fig. 5.21. As soon as the short-circuit, or overload, is removed the regulated voltage regains its former level.

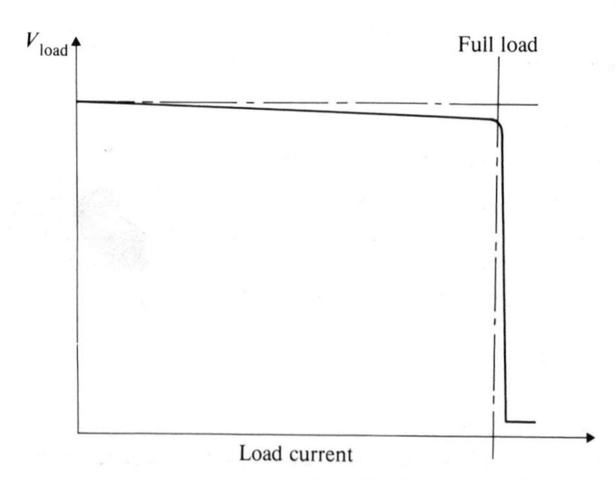

Fig. 5.21 Characteristic of stabilized power supply.

Practical Exercise 5h

Stabilization against load current changes

Connect up the circuit arrangement shown in Fig. 5.22. Vary the value of the load resistance from maximum (5 kΩ) to minimum (100 Ω) and record the values of unstabilized voltage, load (stabilized) voltage, and Zener current for a number of load current values:

Unstabilized voltage (V)________.

Stabilized load voltage (V)________.

Zener current (mA)________.

Load current (mA)________.

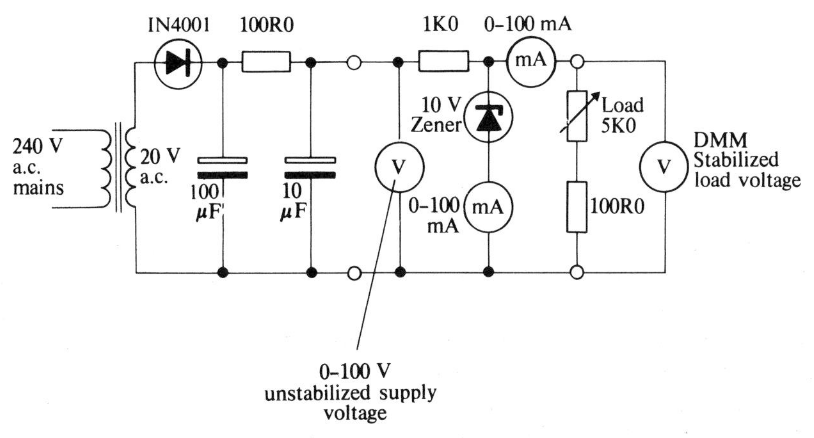

Fig. 5.22 Stabilization against changes in load current.

From these results, note the effect of the change in load current on the stabilized voltage. In addition, note the effect on the Zener current.

Practical Exercise 5i

Stabilization against supply voltage variations

Connect up the circuit arrangement shown in Fig. 5.23. Vary the value of the 5K0 variable resistance (this simulates variations of supply voltage) from maximum (5 kΩ) to minimum (100 Ω) and record values of unstabilized voltage, load (stabilized) voltage, and Zener current for a number of settings of the variable resistance.

Unstabilized voltage (V)________.

Stabilized (load) voltage (V)________.

Zener current (mA)________.

From these results, note the effect of the supply voltage variations on the stabilized voltage.

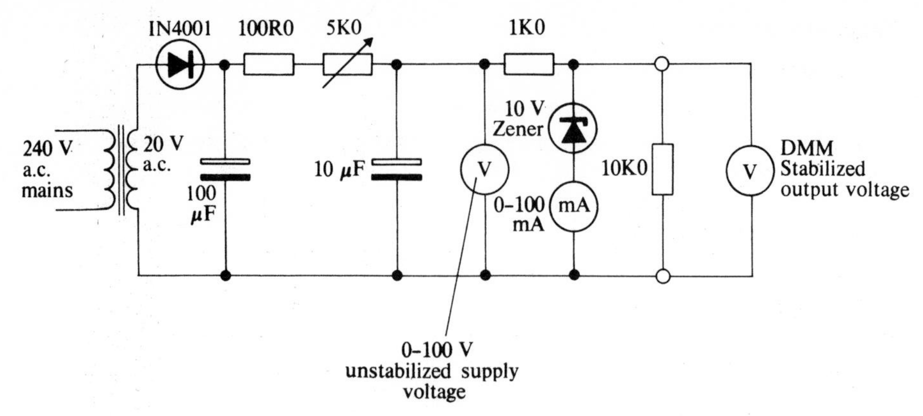

Fig. 5.23 Stabilization against variations in supply voltage.

Practical Exercise 5j

Shunt regulator

Connect up the circuit shown in Fig. 5.24. Vary the value of the load resistance and record values of unstabilized voltage, stablized voltage, and load current for a range of settings.

Unstabilized voltage V_1							
Stabilized voltage, V_2							
Load current, I_1							

From these results, note the effect of load variation on the stabilized voltage.

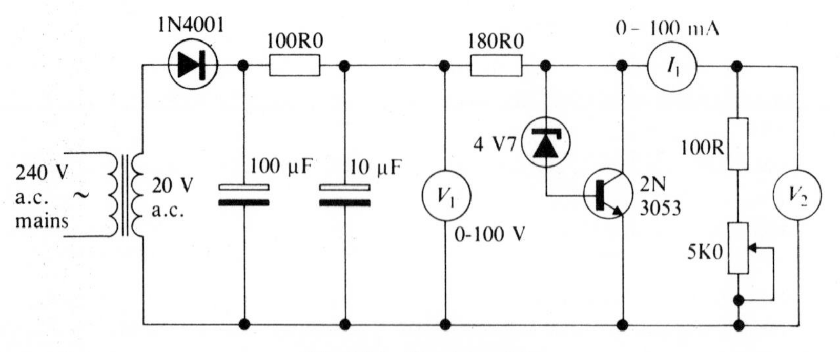

Fig. 5.24 Shunt regulator.

Practical Exercise 5k

Series regulator

Connect up the circuit shown in Fig. 5.25. Vary the value of the load resistance and record values of unstabilized voltage, stabilized voltage, and load current for a range of settings.

Unstabilized voltage V_1								
Stabilized voltage, V_2								
Load current, I_1								

From these results, note the effect of load variation on the stabilized voltage.

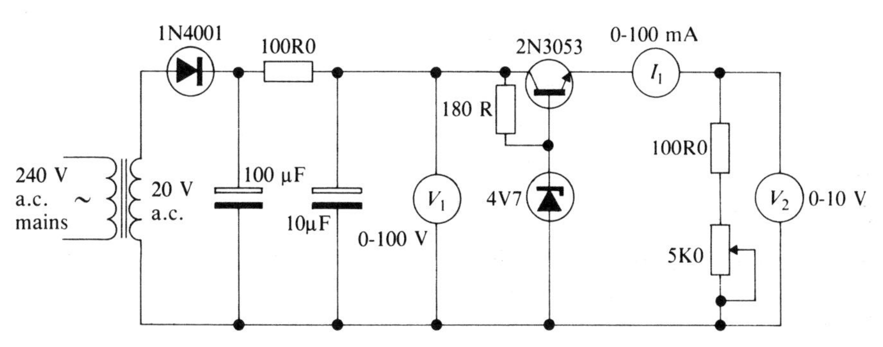

Fig. 5.25 Series regulator.

Practical Exercise 5l

Variable-voltage series regulator

Connect up the circuit shown in Fig. 5.26. Set output voltage by adjusting VR_1.

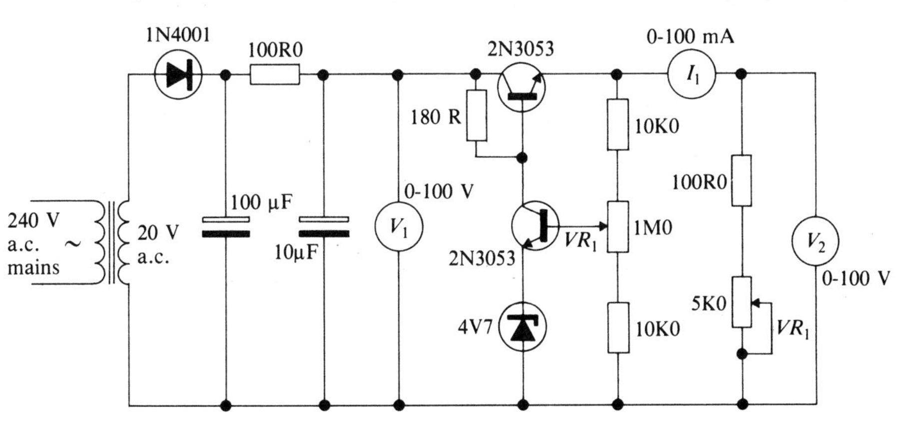

Fig. 5.26 Variable-voltage series regulator.

Record values of stabilized voltage as the load is varied. Repeat for different output voltage settings.

Practical Exercise 5m

Transistor radio power supply test

Connect up the simple Zener shunt regulator power supply shown in Fig. 5.27, with the 9 V transistor radio as the load.

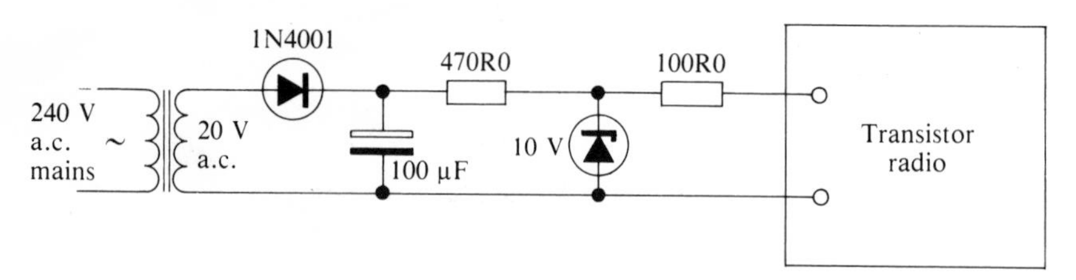

Fig. 5.27 Transistor radio power supply test.

Switch on the power supply, tune-in the radio, and note any distortion that occurs.

Repeat this test, but using the circuit shown in Fig. 5.26 – where the transistor radio replaces the load resistance (5K0 variable resistance).

Compare the results of this test with those of Practical Exercise 5c.

5.12 Clipping

A further application of Zener diodes is that of producing *square* waves by *clipping* a.c. and rectified a.c. waveforms as shown in Fig. 5.28 and Fig. 5.29. The calibration waveform used in some oscilloscopes is often derived in this way.

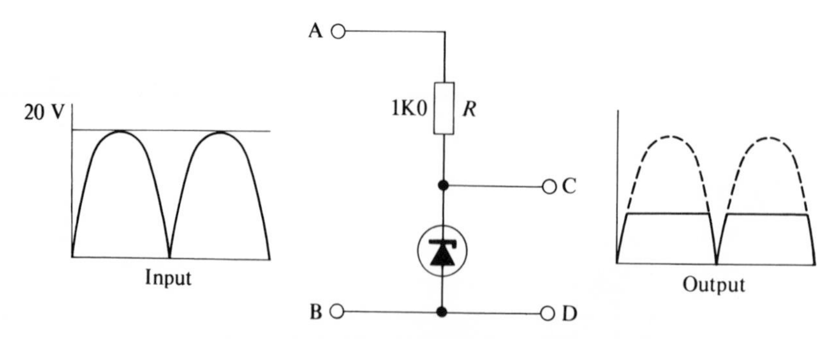

Fig. 5.28 Clipping of full-wave rectified supply.

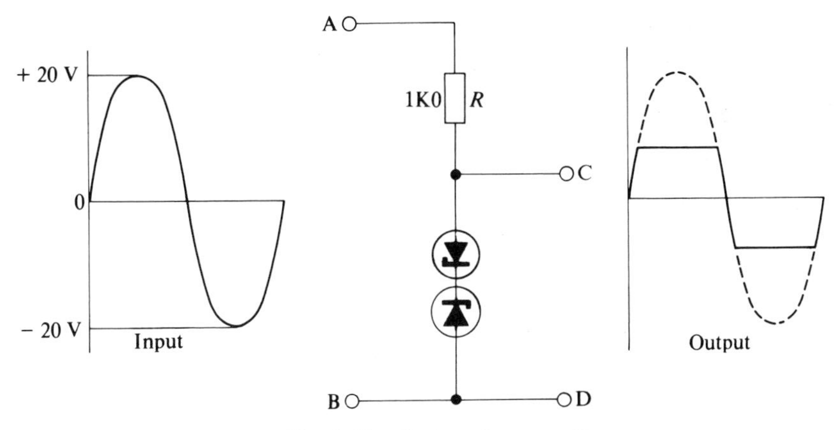

Fig. 5.29 Clipping of a.c. waveform.

Practical Exercise 5n

Simple waveform clipping

Connect up the circuit arrangement shown in Fig. 5.28 and observe the waveforms across AB and CD using the oscilloscope. Repeat for a range of Zener diodes having different reference voltages.

Connect up the circuit arrangement shown in Fig. 5.29 with two Zener diodes connected back-to-back. Observe the waveforms across AB and CD using the oscilloscope. Repeat for a range of Zener diodes.

6. Transistors

6.1 Introduction

The transistor is a three-terminal device as shown in the circuit symbol in Fig. 6.1. After doping of the basic semiconductor material, n and p type semiconductor material is formed and, although many processes of manufacture are used, a transistor is basically a three-layer sandwich of the two types, i.e., n p n or p n p.

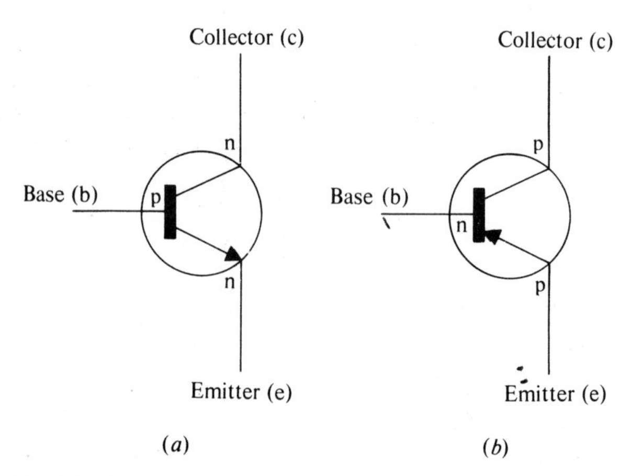

Fig. 6.1 Circuit symbols for the transistor. (*a*) n p n; (*b*) p n p.

The circuit symbols of the two types are similar in appearance, the difference being in the direction of the arrowhead on the emitter lead. As we have seen before this arrowhead indicates the direction of *conventional* current flow which is in opposite directions in the two types, but is *always from the p type to the n type material* in the base-emitter circuit. To avoid the possibility of confusion, we shall consider the n p n type transistor only, except to say that in circuits using p n p type transistors the supply voltage is of opposite polarity and currents flow in opposite directions to those in circuits using n p n type transistors.

6.2 The n p n transistor

The collector and emitter are both n type material and the thin layer between them is the p type base. At first it might be thought that the transistor should work either way, i.e., by interchanging the collector and emitter leads since they are both the same type of material. However, this is not possible, since these regions are not the same size. The collector is larger and in many transistors is directly connected to the case for heat-sinking purposes. During use, most of the heat generated is at the collector–base junction, which must be capable of dissipating this heat. The emitter–base junction can only withstand a small reverse voltage.

Operation in the reverse direction is possible at a very reduced efficiency, but this is not a practical method of connection, since it would very often result in the destruction of the device.

The transistor is generally considered as a current-operated device. If current is passed into the base and through the base–emitter junction, then a positive supply on the collector will cause a current to flow between collector and emitter. Two things should be noted about this *collector* current:

1. For zero base current, the collector current is reduced to *leakage* current, which is less than 1μA under normal conditions (for silicon transistors).
2. For a given base current the collector current which flows is much larger than the base current. This *current gain* is called h_{FE}, where

$$h_{FE} = \frac{i_C}{i_B} = \frac{\text{collector current change}}{\text{base current change}} .$$

Practical Exercise 6a

Static testing of the transistor

Transistors are packaged in many different ways by the various manufacturers, some of the outlines being shown in Fig. 6.2. Although there are several conventional terminal arrangements it is convenient to have a practical test which will determine the type of transistor and identify its terminals. For this static test it is usual to consider the transistor as two diodes connected back-to-back, as shown in Fig. 6.3.

To carry out this test we can use a multimeter on the $\Omega \div 100$ range, *or* the DMM on 1000 kΩ range. It is important to use the *low voltage battery*, since many transistor base–emitter junctions will only withstand about 5 V reverse bias.

Note: If an analogue multimeter is being used, remember to change the leads over, so that the red lead corresponds to battery positive and the black lead corresponds to battery negative. This procedure is not normally necessary when using a DMM, the polarity of the terminals remains as marked for voltage, current, *and* resistance measurement.

The procedure for the transistor test is as follows:

1. Assume that the transistor is an n p n type.

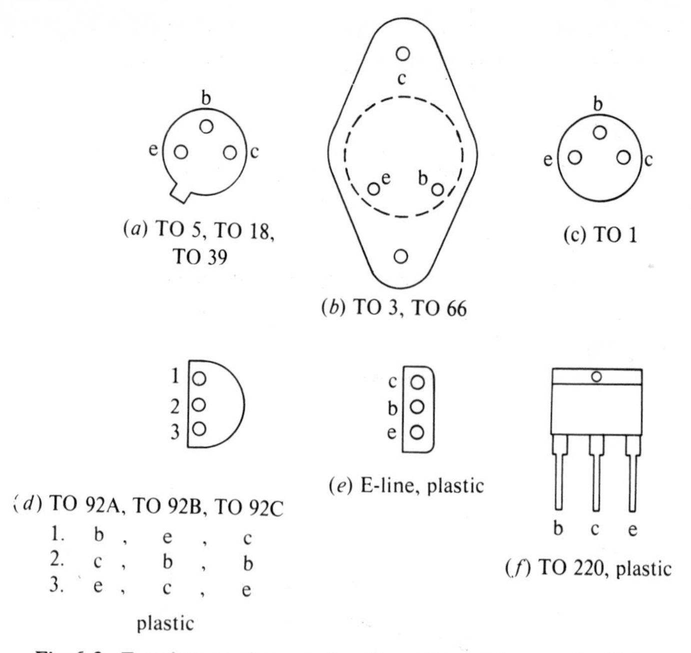

Fig. 6.2 Transistor packages and outlines (viewed from underside).

2. One terminal will conduct to the other two when *positive* is connected to it and *negative* to the other two in turn – this will be the *base* (see Fig. 6.3.)
3. One of the remaining terminals is the collector, the other is the emitter. Connect *positive* to the chosen *collector* and *negative* to the *emitter*. The ohmmeter should show that this is a *high* resistance.
4. Join the collector to the base via a high resistance, e.g. 'wet fingers', and base current will flow causing the collector–emitter resistance to fall. This decrease in resistance is shown on the ohmmeter.

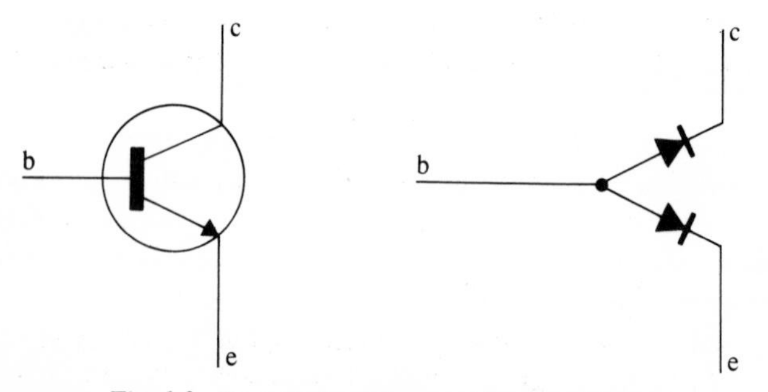

Fig. 6.3 Two-diode analogue of the transistor.

5. If these results are not obtained, the collector must be the lead formerly chosen as the emitter. Reverse the connections and try again!

6.3 Transistor ratings

It is important for every transistor that the ratings quoted in their specifications are not exceeded. Data sheets give the important values, some of which are noted below, and shown in Fig. 6.4.

V_{CBO} = the maximum collector–base voltage (collector + ve)
V_{CEO} = the maximum collector–emitter voltage (collector + ve)
V_{EBO} = the maximum emitter–base voltage (emitter + ve)
P_{tot} = the total power dissipated in the transistor.

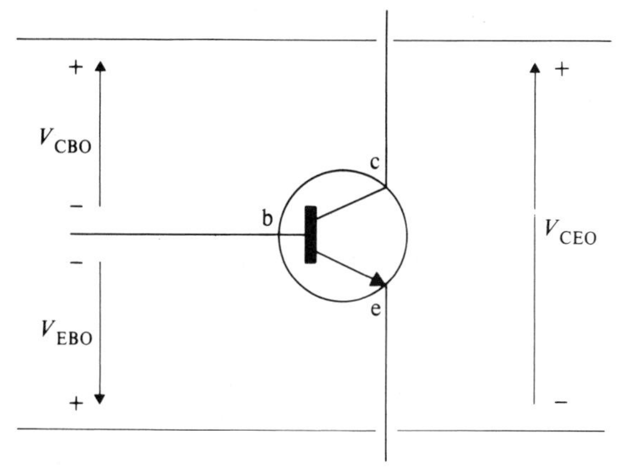

Fig. 6.4 Transistor voltage ratings.

6.4 The transistor as a switch

Consider the transistor in the circuit shown in Fig. 6.5. If the base current I_B was zero, then the collector current I_C would be a small leakage current, and the voltage drop across the load resistor, R_L, would be negligible. Hence,

$$V_{CE} \approx V_{CC} \quad \text{the supply voltage.}$$

If I_B is a small nominal amount then I_C would be equal to $h_{FE}I_B$ and the voltage drop across R_L would be:

$$V_R = I_C R_L \quad \text{and} \quad V_{CE} = V_{CC} - I_C R_L.$$

Increasing I_B would cause I_C to increase further, but a point must be reached when $I_C R_L \approx V_{CC}$, i.e., when I_C cannot be increased any more, regardless of any further increase of I_B.

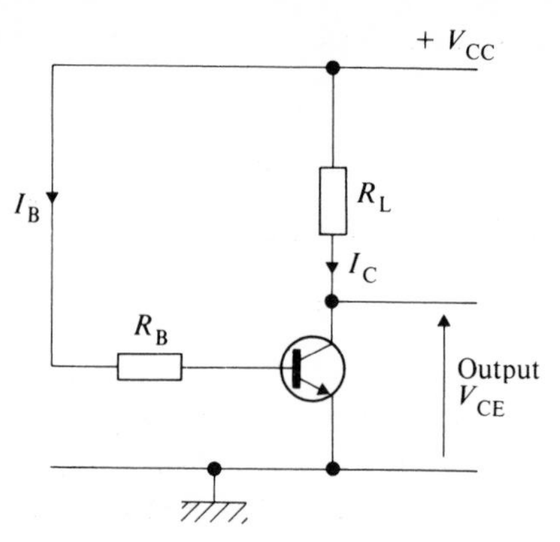

Fig. 6.5 The transistor as a switch.

At this point, the transistor is said to be switched *hard on, bottomed*, or *saturated*, and the voltage V_{CE} is called $V_{CE\ sat}$, the saturation output voltage. This is typically 0.2 V for a silicon transistor and may be as low as a few tens of millivolts but not usually greater than 0.3 V.

It is interesting to note at this point that we can switch much greater load power than the power rating of the transistor – *providing that we switch over quickly*.

EXAMPLE

Consider a transistor switching a 0.5 A load from a 12 V d.c. supply.

1. ***When the transistor is OFF:***

$I_C = 1\ \mu A$ (say), i.e., leakage current only.

$$V_{CE} \approx V_{CC} = 12\ V.$$

Therefore, transistor power dissipation,

$$P = V_{CE} \times I_C$$
$$= 12 \times 1 = 12\ \mu W.$$

2. ***When the transistor is ON (i.e., saturated):***

$$I_C = 0.5\ A,$$
$$V_{CE} = V_{CE\,sat} \approx 0.2\ V.$$

Therefore, transistor power dissipation,

$$P = V_{CE} \times I_C$$
$$= 0.2 \times 0.5 = 0.1\ W.$$

3. ***When the transistor is mid-way:***

$$I_C = 0.25 \text{ A},$$

$$V_{CE} = 6 \text{ V}.$$

Therefore, transistor power dissipation,

$$P = V_{CE} \times I_C$$

$$= 6 \times 0.25 = 1.5 \text{ W},$$

which may be excessive for the transistor.

Providing the high power dissipation area in the middle can be crossed in a short time, the transistor can cope quite happily with the low power ON and low power OFF extremes, and all is well!

Note: The load current must not exceed $I_{C\ max}$.

General purpose transistors switch on and off at hundreds of kilohertz, i.e., on once and off once in less than 10 μs. *High speed* switching transistors are capable of switching at frequencies up to several gigahertz.

Practical Exercise 6b

Simple transistor switching

1. Connect up the circuit shown in Fig. 6.6, and measure the values of V_{BE}, V_{CE}, and I_C:
   (a) with the link OUT,
   (b) with the link IN.
   and complete table below:

	Link OUT	Link IN
V_{BE}		
V_{CE}		
I_C		

2. Now connect up the circuit shown in Fig. 6.7, and set the function generator to 5 Hz square wave with minimum amplitude, and connect it to the signal input terminals of the circuit.

   Note: The purpose of the diode is to protect the base-emitter junction from reverse voltages during the negative half cycles of the applied signal.

   (a) Increase the amplitude of the signal until the transistor switches the lamp ON and OFF.
   (b) Use a dual-trace CRO to examine the input square wave and the output voltage waveform V_{CE}, and measure the time taken for the transistor to switch.

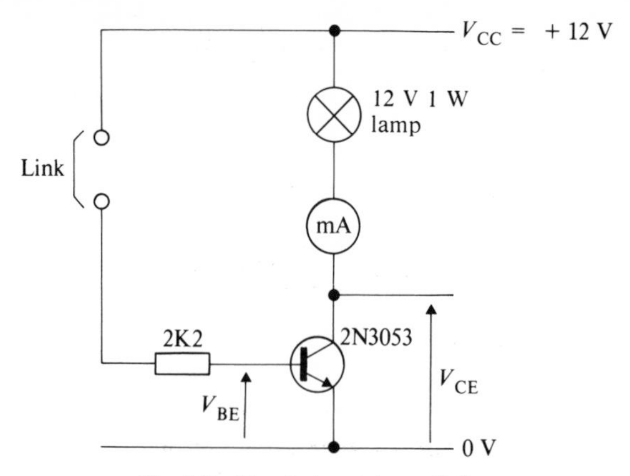

Fig. 6.6 Simple transistor switch.

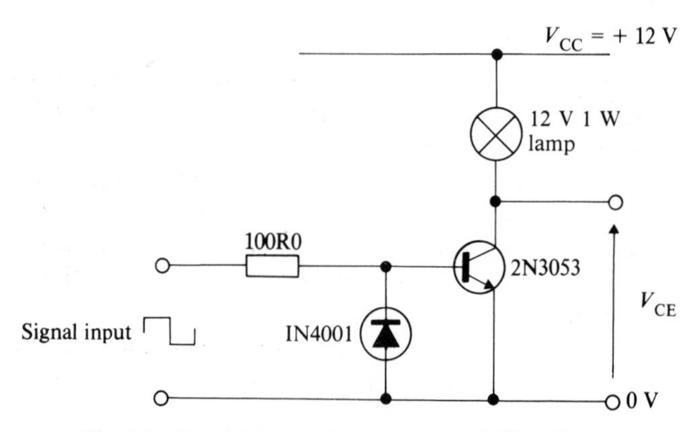

Fig. 6.7 Transistor switch to measure switching time.

(c) Increase the frequency of the signal, and see if this affects the switching time.

Practical Exercise 6c

Light-operated switch

Connect up the circuit shown in Fig. 6.8 and try different values of resistor *R*, until the light comes ON when the ORP12 is shielded from the light, i.e., automatic parking light.

Now, change the positions of *R* and the ORP12 and see what happens.

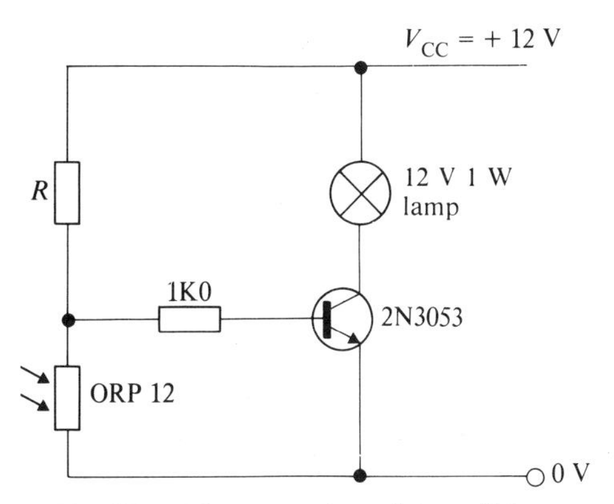

Fig. 6.8 Light-operated transistor switch.

Practical Exercise 6d

Manually-operated bistable switch

Connect up the circuit shown in Fig 6.9, in which the two transistors are said to be cross-coupled. When the supply is switched on, only *one* lamp will light. This is because the transistor which turns on first will have a saturation voltage of only about 0.2 V which is not enough to turn the other transistor on. From Practical Exercise 6b you found that V_{BE} was about 0.7 V for ON condition.

Now, take a wander lead from the common 0 V and 'touch' the other end to the base input terminal of the ON transistor. What happens? Now touch the wander lead to the other base input terminal.

As you can see the circuit has *two* stable states with one transistor ON and the other OFF. This type of circuit is called a *bistable*, and is the basic element

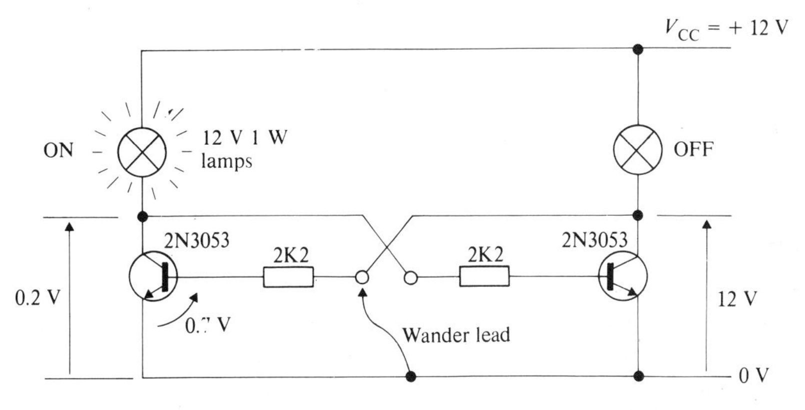

Fig. 6.9 Manually-operated transistor bistable switch.

of counting circuits. Other names for the bistable are: memory, latch, flip-flop, and divide-by-two depending on the application.

6.5 The transistor as an amplifier

Many different types of amplifier are used in the field of industrial electronics, most of which can be made using transistors. At this stage, we are going to examine a simple a.c. voltage amplifier, in which a small a.c. signal is applied to a pair of input terminals, and the amplifier is required to reproduce a linearly amplified version of the applied signal at a pair of output terminals. A further requirement is that the amplifier must be capable of performing its function over a range of frequencies of the input signal – this range being termed the *bandwidth* of the amplifier.

The circuit arrangement we shall consider is referred to as the *common emitter configuration* in which the emitter is common to both the input and output circuits. The input is applied between the base and emitter, and the output is taken from the collector and emitter.

6.6 Biasing

Before feeding an a.c. signal into the base–emitter circuit we must set V_{CE}, by appropriate biasing, so that we shall get an undistorted output. For this, we set V_{CE} to a value approximately mid-way between V_{CC} and zero. Now the increase and decrease of I_C (due to the signal causing changes in I_B) can cause an equal decrease and increase of V_{out} without distortion.

Various methods of biasing have been used, the common arrangements are shown in Fig. 6.10. Consider the simple bias circuit shown in Fig. 6.10(*a*), the basic object is to feed in a steady base current, I_B, so that with no signal applied (the *quiescent* state) this base current is sufficient to cause a steady collector current, I_C, to flow such that $I_C R_L \approx \frac{1}{2} V_{CC}$.

The signal – usually an a.c. sine wave – can now be fed in via capacitor C_1, which is used to block the passage of direct current, and, therefore, ensure that only the a.c. signal is passed in to the transistor for amplification.

The signal voltage causes the base current to vary sinusoidally above and below the steady bias current I_B, which in turn causes the collector current to vary sinusoidally above and below the quiescent value I_C. Since this changing collector current flows in the load resistor, R_L, then a sinusoidal a.c. voltage is developed across R_L which appears at the collector. This a.c. voltage is passed out via capacitor C_2, so that it is only the a.c. voltage that is passed on, the d.c. level (V_{CE}) is blocked by C_2. The voltage gain of the amplifier circuit is given by

$$\text{Voltage gain} = \frac{V_{out}}{V_{in}} \quad (\text{and is not} = h_{FE}).$$

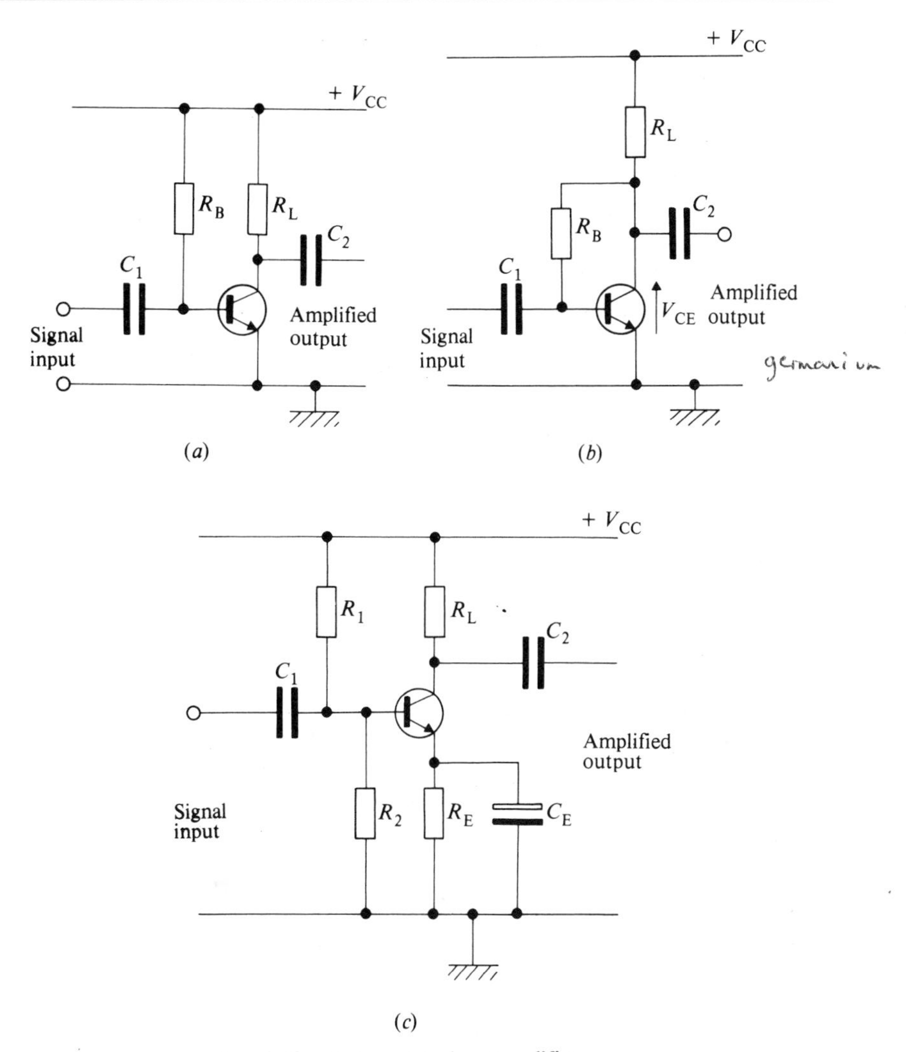

Fig. 6.10 Biasing a transistor amplifier.

Practical Exercise 6e

Simple transistor voltage amplifier

Connect up the circuit shown in Fig. 6.11. Switch on the supply voltage:

1. Adjust the 1M0 variable resistor to give $V_{CE} = +6$ V.
2. Connect a sine wave signal of a few hundred millivolts at 10 kHz to the signal input terminals.
3. Use a dual-trace CRO to measure the input and output voltages, and work

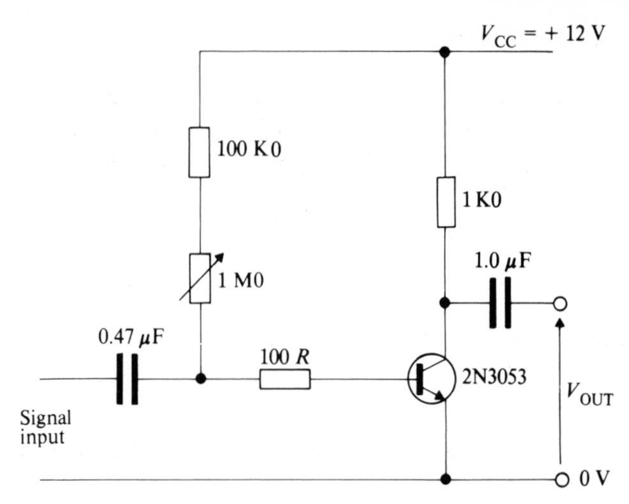

Fig. 6.11 Simple transistor voltage amplifier.

out the voltage gain. Also note the 180° phase difference between the input and output waveforms.

4. Increase the amplitude of the signal and observe the effect on the output waveform.
5. Remove the signal input and reset the bias to give V_{CE} = +9 V, and repeat the above procedure. Note which part of the output waveform distorts first when the input signal is varied. Why?
6. Repeat, with the bias set to give V_{CE} = + 3 V.
7. Reset the bias to give V_{CE} = + 6 V, and adjust the signal amplitude to produce an undistorted output voltage waveform. Now, decrease the frequency of the signal from 10 KHz and find the frequency f_1 at which the amplitude of the output waveform falls to 0.7 of its value at 10 kHz. f_1 = ________ Hz.
8. Increase the frequency of the signal above 10 kHz and find the frequency f_2 at which the amplitude of the output waveform falls to 0.7 of its value at 10 kHz. f_2 = ________ Hz.

The output voltages at f_1 and f_2 are known as the *half power* or *–3 dB points*, and define the useful frequency range of operation of the a.c. amplifier, i.e., its *bandwidth.*

$$\text{Bandwidth} = f_2 - f_1 = \_\_\_\_\_\_\_\_ \text{Hz.}$$

6.7 Field effect transistor (FET)

The circuit symbols for field effect transistors (FETs) are shown in Fig. 6.12. FETs are not current-operated devices, but depend for their operation on the

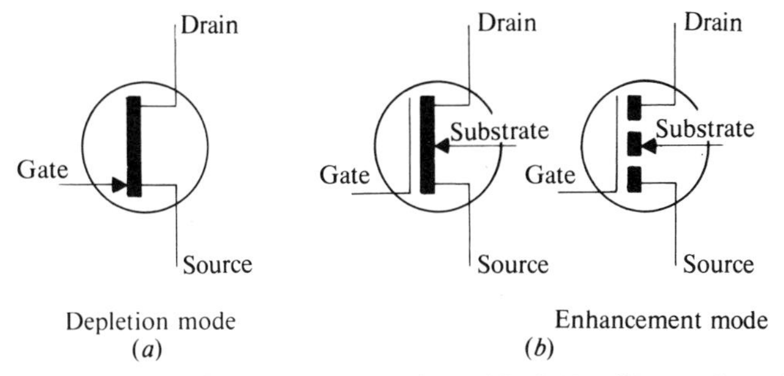

Fig. 6.12 Circuit symbols for FETs. (*a*) An n channel JUGFET; (*b*) an n channel MOSFET (insulated gate-IGFET).

effect of an electric field produced by the application of an input voltage to the gate terminal. This electric field controls the width of the *channel* in which conduction takes place between the drain and the source. FETs are, therefore, effectively voltage-operated devices and have a very high input impedance, typically several megohms, and high gain, but have not yet been developed to handle reasonable power levels.

6.8 The FET as a switch

Consider the circuits shown in Fig. 6.13. To switch the FET on, V_{GS} should be zero, so that a negative pulse must be applied to the diode to switch off the FET in the series circuit of Fig. 6.13(*a*). In the shunt circuit shown in Fig. 6.13(*b*), the application of a negative pulse to the gate will switch off the FET, thus allowing the signal through to the load.

Practical Exercise 6f

FET as a switch

Connect up the circuit as shown in Fig. 6.13(*a*).

Apply a signal to the input (V_S) and monitor the output voltage using the CRO when a pulse input is applied to the diode. Record the waveforms of input and output.

Repeat this exercise with the circuit shown in Fig. 6.13(*b*).

Practical Exercise 6g

The FET amplifier

Connect up the circuit shown in Fig. 6.14.

Adjust the 1M0 variable resistor (bias) to give $V_{GS} = 0$ V. Set the function generator to SINE, 1 kHz, 4 V peak-to-peak and connect it to the input terminals AB. Using a dual-trace CRO, measure the input signal (AB) and the output

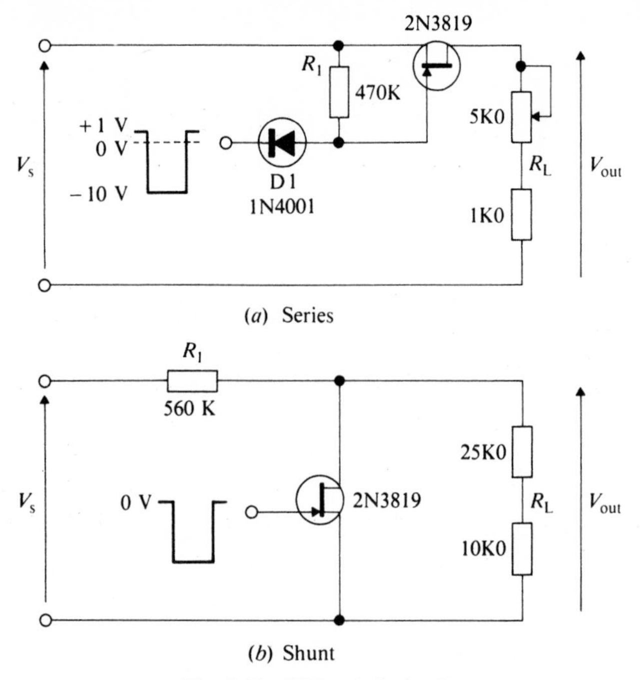

(*a*) Series

(*b*) Shunt

Fig. 6.13 FET switch circuits.

signal (CB). Record the amplitude (and phase) of the input and output signals. What is the voltage gain?

Disconnect the lead from the gate terminal of the FET and observe the effect on the amplitude of the input signal. What does this tell us about the input impedance of the FET?

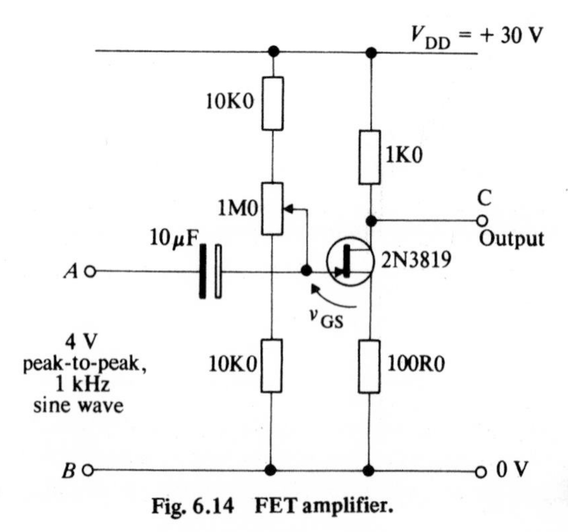

Fig. 6.14 FET amplifier.

7. Unijunction Transistors (UJT)

7.1 Introduction

The unijunction transistor (UJT) is a three-terminal semiconductor device which is packaged in a similar way to ordinary transistors, e.g., TO can or plastic package, but whose action is distinctly different. The circuit symbol is shown in Fig 7.1, and it can be seen that this is different to the symbol of the ordinary transistor. The terminals B_2 and B_1 refer to *base 2* and *base 1* whilst E indicates the *emitter.* The base 1 to base 2 connection is in the form of a silicon bar which has a resistance of between 5 kΩ and 10 kΩ either way round under static conditions. The junction between the emitter and base 1 is unidirectional as indicated by the arrowhead on the symbol.

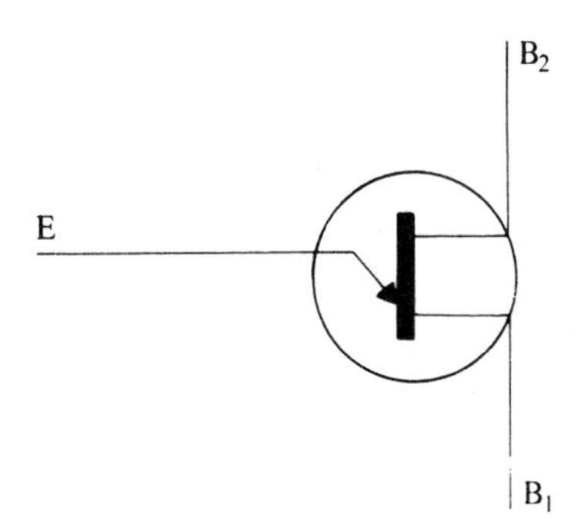

Fig. 7.1 Circuit symbol for the unijunction transistor.

7.2 Action of the UJT

The supply voltage applied to the UJT between base 2 and base 1, i.e., V_{BB} will cause a small nominal current to flow. It will be found that the voltage between the emitter and base 1 will always be a fixed proportion of V_{BB}, and is given by:

$$V_{EB_1} = \eta V_{BB},$$

where η is the *intrinsic standoff ratio.*

The intrinsic standoff ratio can be between about 0.5 and 0.8. As we shall see later, 0.7 is a very useful value of η for the UJT to have.

If the voltage V_{EB1} is forced up by some external circuit arrangement, a value V_P will be reached which is sufficient to turn on the emitter–base 1 junction, i.e., it will suddenly become a low resistance, and a large emitter current I_E can flow.

A common circuit arrangement to apply the appropriate voltage to the emitter–base 1 junction is the *CR* charging network. The voltage across the capacitor rises as the capacitor charges and reaches a value of about 0.63 V_S volts after a time of *CR* seconds. Since V_P for the UJT is linked to the η value, then we can arrange for the UJT to 'fire' after *CR* seconds (the *time constant*) – a very useful system.

6.3 The UJT as a relaxation oscillator

Consider the circuit arrangement shown in Fig. 7.2. When the supply voltage V_{BB} is switched on, the UJT is essentially OFF. Capacitor C commences its charge through R – the rate being dependent upon the values of C and R. When v_C reaches the value of V_P, the emitter–base 1 junction is turned on, and the charge stored in the capacitor can rush out through the emitter–base 1 junction and resistor R_{B1}. This causes *two* things to happen:

1. the current pulse through R_{B1} develops a positive output pulse of voltage across R_{B1};
2. the current flowing between base 2 and base 1 increases, and the voltage drop across R_{B2} increases, which causes a negative-going pulse from the + V_{BB} level and is obtained at output 2, i.e., between base 2 and 0 V.

As the charge on the capacitor diminishes rapidly, the voltage is no longer able to keep the emitter–base 1 junction turned on, and it reverts to its non-

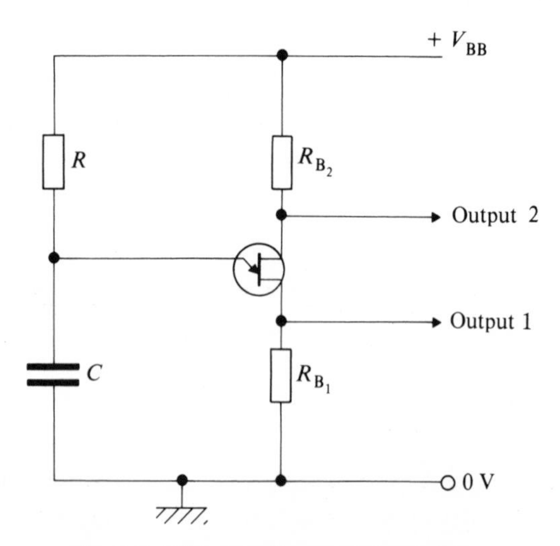

Fig. 7.2 UJT relaxation oscillator circuit.

conducting state. The voltage at output 1 falls to zero and the voltage at output 2 returns to + V_{BB}.

The capacitor charges from this residual level and the cycle repeats itself indefinitely – hence the name *relaxation oscillator.* This type of circuit is self-starting, i.e., it commences oscillating as soon as the supply is switched on. The waveforms for this circuit are shown in Fig. 7.3.

The UJT relaxation oscillator is used to supply pulses to trigger stroboscopes, thyristors, triacs, and many other circuits.

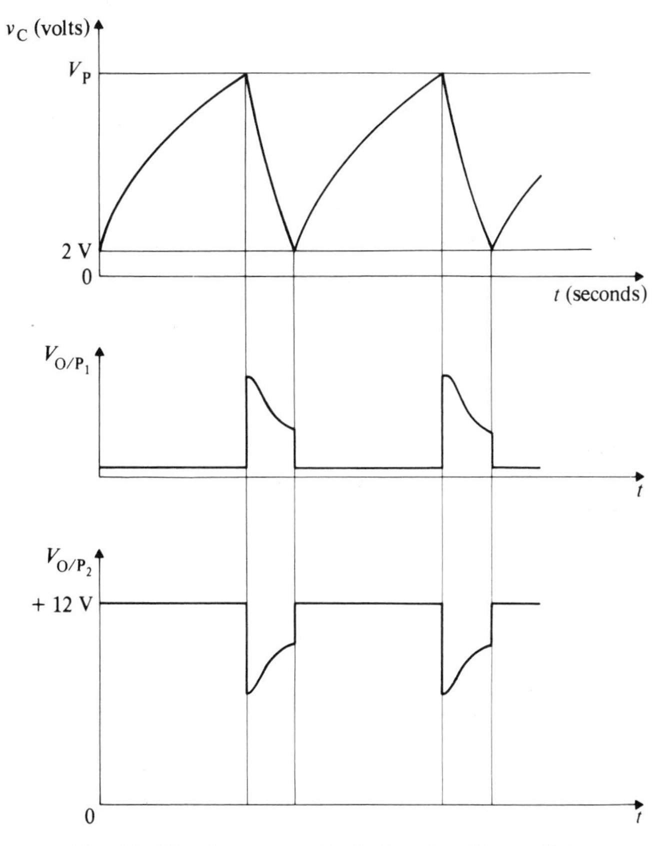

Fig. 7.3 Waveforms occurring in the relaxation oscillator.

Practical Exercise 7a

Intrinsic standoff ratio in UJT

Connect up the circuit arrangement shown in Fig. 7.4. Increase the voltage V_{EB1} until the UJT is triggered, and note the value of V_{EB1} at which this occurs. This is the value V_P.

The intrinsic standoff ratio η is given by

$$\eta = \frac{V_{EB1}}{V_{BB}} = \_\_\_\_\_\_\_ .$$

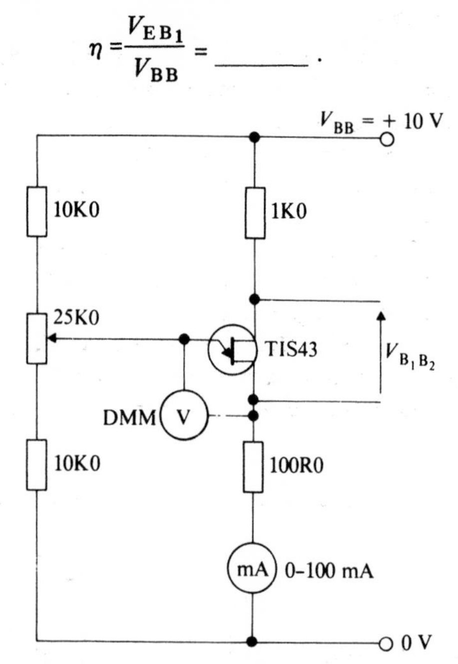

Fig. 7.4 Intrinsic standoff ratio.

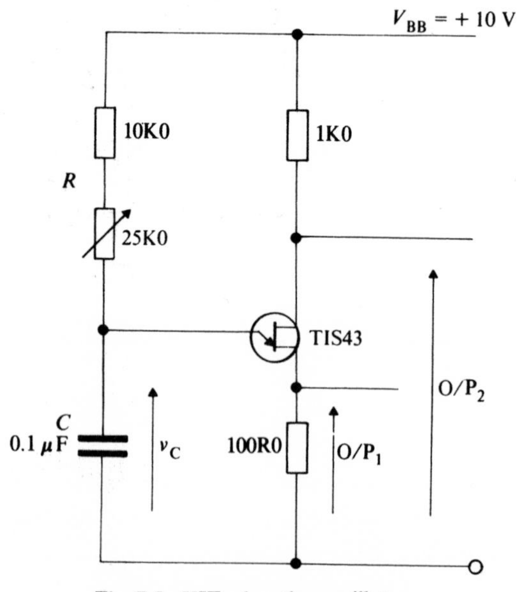

Fig. 7.5 UJT relaxation oscillator.

Practical Exercise 7b

UJT relaxation oscillator

Connect up the circuit shown in Fig. 7.5. Connect the inputs of a dual-trace CRO across the capacitor C to examine the waveform v_C, and across the 100R0 resistor to examine the voltage waveform of output 1.

Do the waveforms observed compare to those shown in Fig. 7.3?

Now, observe the voltage waveform of output 2 and check that this corresponds to the waveform shown in Fig. 7.3.

Revert back to observing the voltage waveforms of v_C and output 1, and try changing the values of C and R, noting the effects on the waveforms – particularly their size, duration, and frequency.

This oscillator can also be used on rectified a.c. supplies (clipped if necessary), and is commonly used in this manner when supplying pulses to thyristors, as shown in Fig. 7.6. The pulses are now *synchronized* to the a.c. supply.

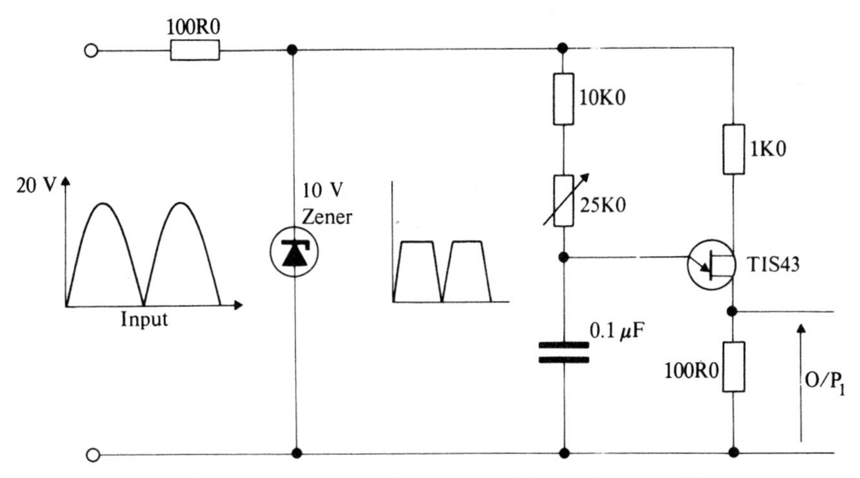

Fig. 7.6 UJT relaxation oscillator supplied from clipped rectified a.c.

8. Thyristors and Triacs

8.1 Introduction

The circuit symbol for the thyristor is shown in Fig. 8.1, and it appears to be like a diode except that it has an extra terminal called the *gate.* A diode will conduct when forward biased but not when reverse biased, i.e., it will only conduct when the anode is positive with respect to the cathode. The thyristor will not conduct when reverse biased and in this respect is similar to the diode. We find, however, that the thyristor will not conduct in the forward direction unless a specified current is passed into the gate, i.e., we can control when the device conducts. This led to the thyristor being originally refered to as a silicon controlled rectifier (SCR).

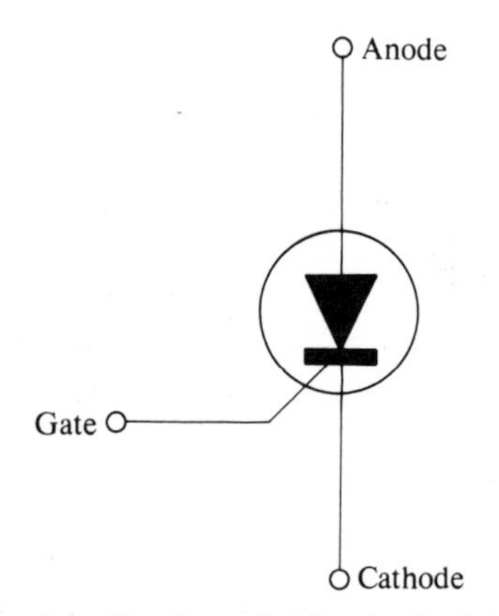

Fig. 8.1 Circuit symbol for the thyristor.

Consider the thyristor characteristic shown in Fig. 8.2. The reverse characteristic is similar to that of a diode. When forward biased the forward voltage can be increased until at a value V_{BO}, the *breakover voltage*, the device conducts and the characteristic becomes much the same as that of the diode, except that $V_F \approx 1$ V (not 0.7 V). The current at breakover is called the *latching current.* If gate current is passed in, this has the effect of reducing V_{BO} until if enough current is passed in the characteristic is virtually the same as that of a diode.

Once the thyristor is on it is self-sustaining, i.e., it *latches on*, and we can remove the gate current.

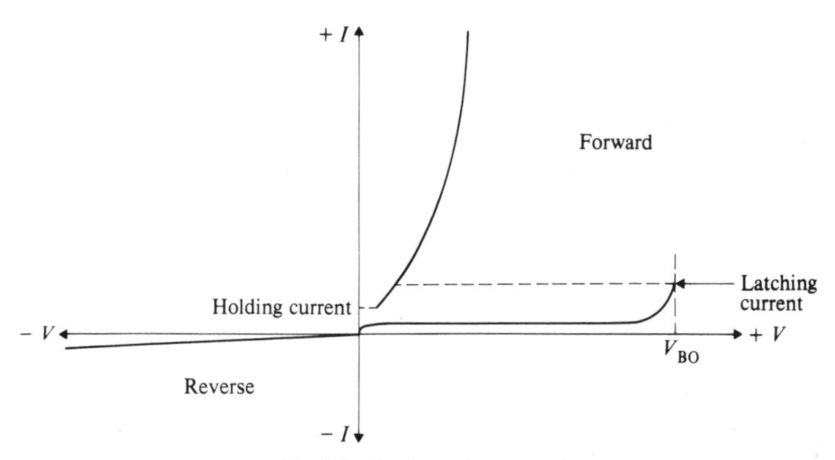

Fig. 8.2 Thyristor characteristic.

8.2 Methods of turning thyristors ON (firing or triggering)

Several methods are available for turning a thyristor on. We shall consider these in turn:

1. ***By raising the forward voltage above V_{BO}.*** Not used.
2. ***By raising the forward voltage at too fast a rate.*** This is indicated by the dV/dt rating of the thyristor. Practical triggering arrangements do not use this. The dV/dt rating is an indication of the maximum rate of rise of transient on the supply before spurious firing occurs.
3. ***By supplying current to the gate.*** This will be as a result of supplying a voltage of the correct polarity, i.e., positive for gate input current, and more than the minimum size for more than the minimum time. Figures stating these minimum values are found in manufacturer's data and are used to determine the minimum height and width of voltage pulse needed to trigger the thyristor. This current may be supplied:
   (a) ***Direct from the supply positive via a resistor.*** This is possible but wasteful since it would not be needed after the thyristor is switched on.
   (b) ***From a charged capacitor.*** Arrangements are necessary to recharge the capacitor for the next firing time. An indirect way of doing this is to use the unijunction relaxation oscillator already seen in Chap 7.

8.3 Methods of turning thyristors OFF (commutating)

As we have seen, the thyristor latches on once it is triggered. If operated on a steady d.c. supply we appear to have no way of switching it off.

Consider the following methods:

1. ***By reducing the forward current below the holding value.*** This is a general

requirement. The *holding current* is less than the *latching current* and if we can reduce the forward current below this then the thyristor will revert back to the non-conducting state. We can achieve this by disconnecting the supply, but any use of the thyristor as a switch would be defeated since we need another switch to control it.

2. ***By applying a reverse voltage across the thyristor.*** This can be done by connecting a charged *commutating* capacitor across the thyristor as shown in Fig. 8.3. This achieves two things at the same time. It applies the reverse voltage and also the rapid current flow into the capacitor needed to discharge it, diverts current from the thyristor, and its current falls below the holding value. The thyristor turns off and the capacitor charges to the supply voltage. Again, commutating circuitry is needed to operate this automatically. The capacitor required is very large for heavy currents.

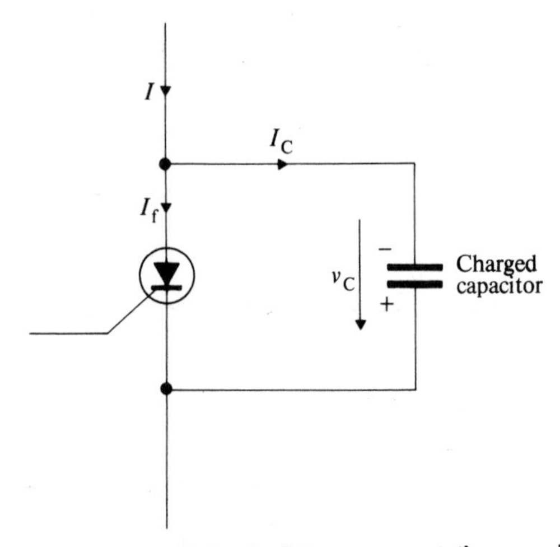

Fig. 8.3 Switching off the thyristor – commutating capacitor.

3. ***By using a rectified supply.*** If the supply is unsmoothed full-wave rectified it will fall to zero twice per cycle, and the thyristor will switch off each time.
4. ***By using the thyristor on the a.c. supply.*** The thyristor will switch off at the end of each positive half-cycle and will not conduct during the negative half-cycles.

8.4 Isolation of the trigger circuit

In many cases thyristors are used in circuits with a.c. mains voltages supplied. The low-voltage trigger circuits need to be isolated from these mains voltages.

Two methods of achieving this isolation are considered below:

1. ***By current transformer***, as shown in Fig. 8.4(*a*).
2. ***By optical isolator***, as shown in Fig. 8.4(*b*). The input is fed into a light-emitting diode (LED) and the light emitted is sensed by a photo transistor. This turns the transistor on and the resulting current pulse generates a voltage across an external resistor which is connected to the thyristor gate/cathode circuit. Isolation between the LED and the photo diode can be up to 2.5 kV.

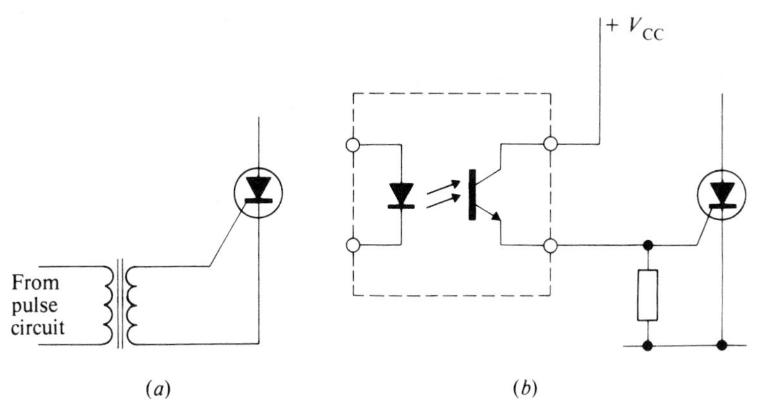

Fig. 8.4 Isolation of trigger circuit. (*a*) Circuit transformer; (*b*) optical isolator.

8.5 Phase control of the thyristor in a.c. circuits

Here, the thyristor is used as a switch to control the current flowing in an a.c. load. The thyristor is switched on by the first suitable pulse during a positive half-cycle and remains on for the rest of that half-cycle. As the supply voltage falls to zero the current is reduced below the holding current level and the thyristor switches off. It remains off during the negative half-cycle. The sequence repeats itself during each cycle, and the waveforms are shown in Fig. 8.5.

The trigger signals can be obtained from a unijunction relaxation oscillator circuit as previously described, and if operated from a full-wave rectified circuit, will be *synchronized* to the a.c. supply of the thyristor so that for a given setting of *CR* the thyristor will always receive its trigger pulse at the same angle. The angle may be varied by changing the value of *CR*. In practice, this is achieved by making *R* a variable resistor.

This method of control is used in speed and light control circuits, in power supply control replacing rectifiers and many other applications. The main disadvantage is that one half-cycle is lost. The second half-cycle can be used if two thyristors are connected back-to-back in parallel, i.e., *inverse-parallel connection* and fired on alternate half-cycles.

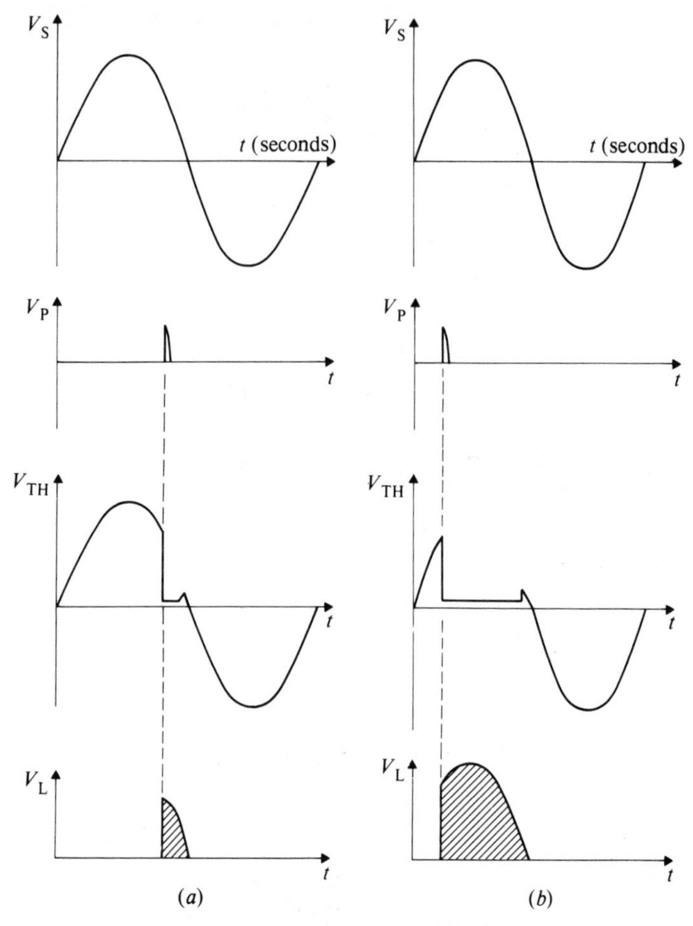

Fig. 8.5 Waveforms with phase control. (*a*) Late firing angle (*b*) early firing angle.

8.6 Burst triggering

Phase control is not used in some applications since switching thyristors on and off during conduction generates a tremendous amount of radio frequency noise. Also, some circuits, like heaters, respond slowly to changes. Here the thyristor can be turned on as near to zero as possible and left on for a time covering a number of complete cycles. Control is effected by varying the number of conducting cycles.

Practical Exercise 8a

Turning ON thyristors

Connect up the circuit shown in Fig. 8.6. Vary the resistance until the lamp

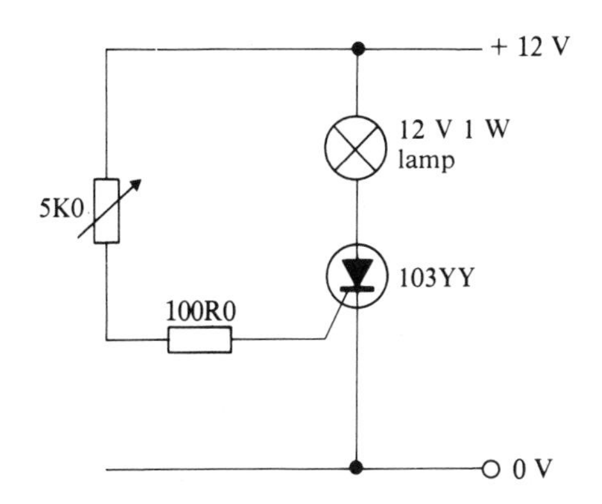

Fig. 8.6 Turning ON the thyristor.

comes on. Then disconnect the gate circuit. Does lamp stay on?

Now try the same circuit with a 24 V lamp and the 20 V a.c. supply.

Practical Exercise 8b

Turning thyristors ON and OFF

Connect up the circuit shown in Fig. 8.7.

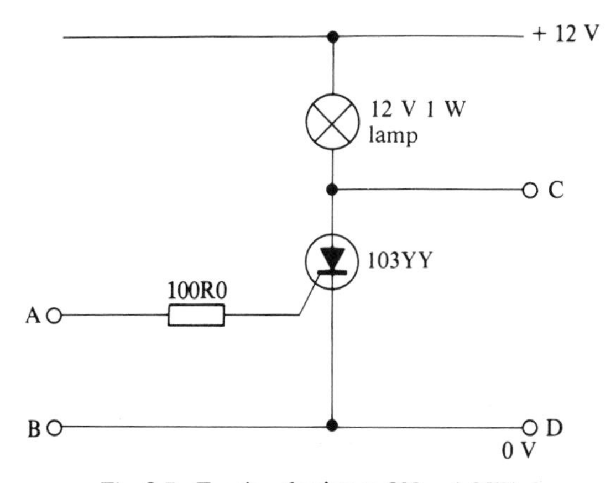

Fig. 8.7 Turning thyristors ON and OFF.

1. Charge various capacitors to + 12 V and connect them across the terminals AB to try to switch the thyristor on. Does it matter which way round the capacitor is connected? Does the capacitor size matter?
2. Now, if you have the lamp ON, try to turn it OFF by placing a charged capacitor in reverse across the terminals CD.

   Do not use electrolytics.

   Does the capacitor size matter?

Practical Exercise 8c

Lamp dimming circuit

Connect up the circuit shown in Fig. 8.8.

Examine the waveforms across the lamp (load) and across the thyristor using a dual-trace CRO. *Use the junction of the lamp (load) and the thyristor as the common earth connection.*

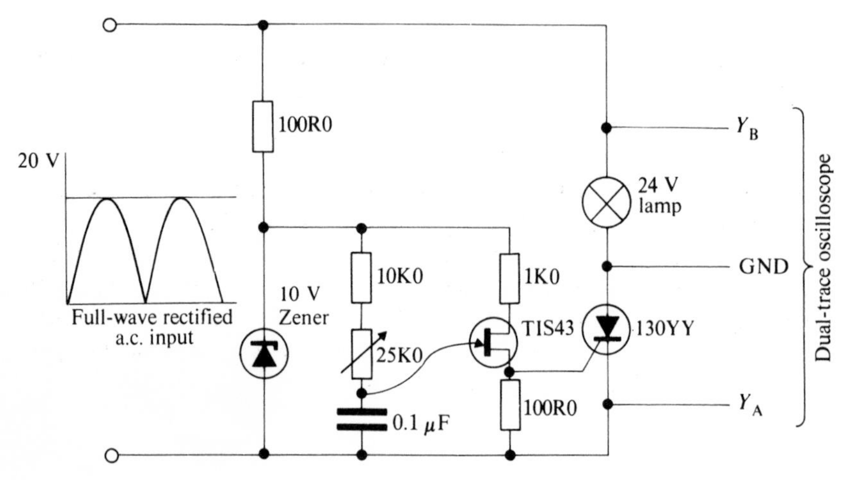

Fig. 8.8 Lamp dimming circuit.

Note: This connection of the CRO will cause the thyristor waveform to be inverted.

Vary the firing angle by adjusting the 25K0 variable resistor and observe the effect on the waveforms, and the brilliance of the lamp.

Connect the oscilloscope leads across the thyristor and the 100R0 resistor. (*Use the low side of the circuit as the common earth this time.*) Vary the firing angle as before, and observe the effects on the waveforms.

Note: You will not be able to get the thyristor to trigger much before 20° and stay on much after 160°. Not to worry, 95 per cent of the power available is present between these limits.

8.7 The triac

The triac was developed as a result of the problems created by the inverse-parallel connection of thyristors and the requirement of separate trigger signals. The circuit symbol is shown in Fig 8.9(*a*) which suggests that it is like two back-to-back thyristors as shown in Fig. 8.9(*b*). The difference is that the triac can be triggered on either half by either a positive or a negative pulse on to its single gate terminal.

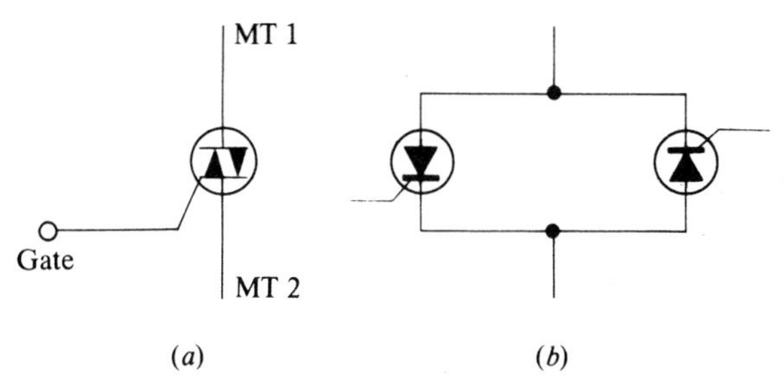

Fig. 8.9 The triac. (*a*) Circuit symbol; (*b*) equivalent circuit of triac (inverse-parallel connection of thyristors).

The triac can be triggered on both half-cycles by the UJT relaxation oscillator, as shown in Fig. 8.10, but this requires a rectified supply. An alternative is to use a *diac*, which is a bi-directional diode as shown in Fig. 8.11(*a*). The diac conducts in either direction if sufficient voltage is applied across it, i.e., it has a breakover voltage for both forward and reverse polarity. Symmetrical and asymmetrical diacs are available, and a typical breakover voltage of the symmetrical type is ± 30 V as shown in Fig. 8.11(*b*).

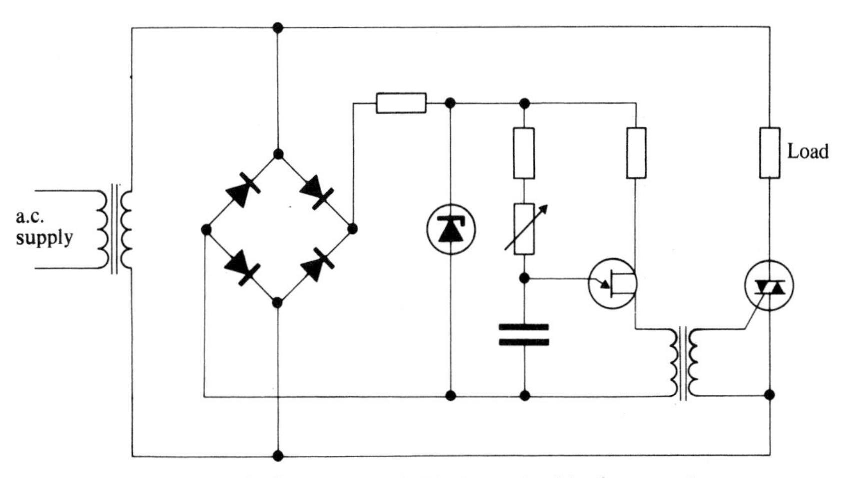

Fig. 8.10 Full-wave (UJT trigger circuit) triac control.

A simple *CR* circuit can be used to control the voltage applied to the diac, which is used to block the gate of the triac until the diac breakover, when a pulse of current flows to trigger the triac as shown in Fig. 8.12.

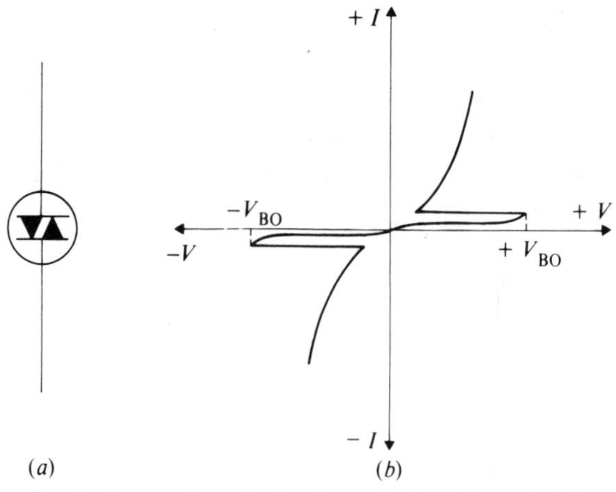

Fig. 8.11 The diac. (*a*) Circuit symbol; (*b*) characteristic.

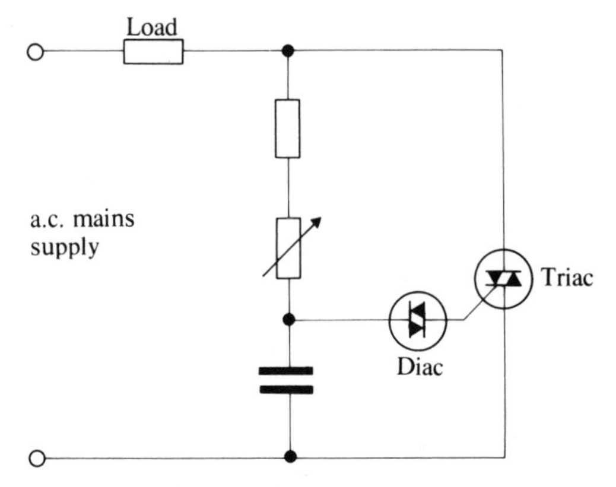

Fig. 8.12 Phase control of triac – using diac triggering.

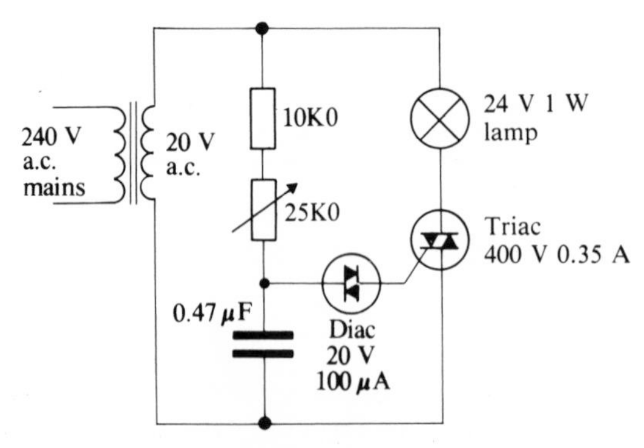

Fig. 8.13 Triac lamp dimming circuit, using diac trigger circuit.

Practical Exercise 8d

Triac control of lamp

Connect up the circuit shown in Fig. 8.13. Observe the waveforms across the triac and the load using a dual-trace CRO.

Adjust the 25K0 variable resistance and note the effect on the waveforms and on the brightness of the lamp.

9. Pulse Shaping

9.1 Introduction

It is necessary to be able to produce many different waveforms for use in electronics. Some of the most commonly used waveforms are as shown in Fig. 9.1. Several waveforms may be obtained by modifying the shape of an existing waveform, e.g., a square may be produced by cutting off the top and bottom of a sine wave – this is known as *limiting* or *clipping*, as shown in Fig. 9.2(*a*). A short

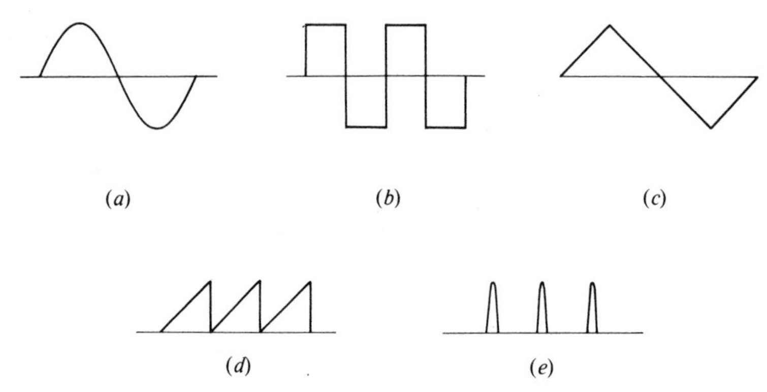

Fig. 9.1 Commonly used waveforms. (*a*) Sine; (*b*) square; (*c*) triangular; (*d*) ramp; (*e*) pulse.

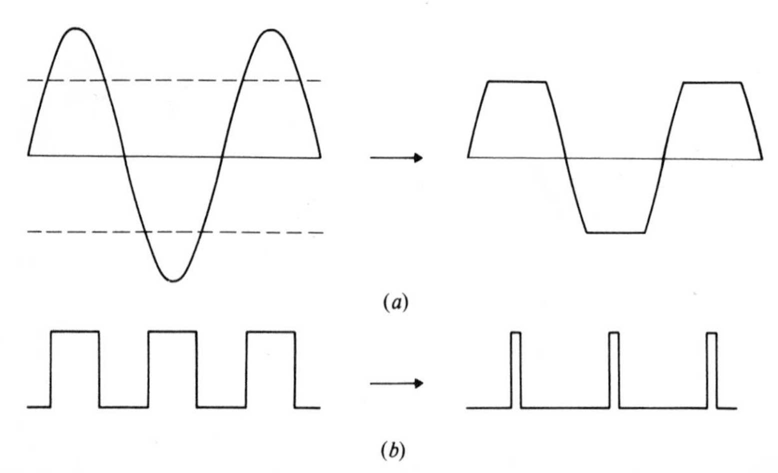

Fig. 9.2 Wave shaping. (*a*) Producing a square wave from a sine wave (limiting or clipping); (*b*) producing short-duration pulses from square waves.

duration pulse waveform is often required to trigger thyristors and triacs, and this may be achieved from a square wave as shown in Fig. 9.2(*b*).

9.2 Waveform generation

1. ***Sine waves*** are generally produced by tuned LC or frequency-selective RC circuits in conjunction with an amplifier, the arrangement being known as an *oscillator*, as shown in Fig. 9.3.

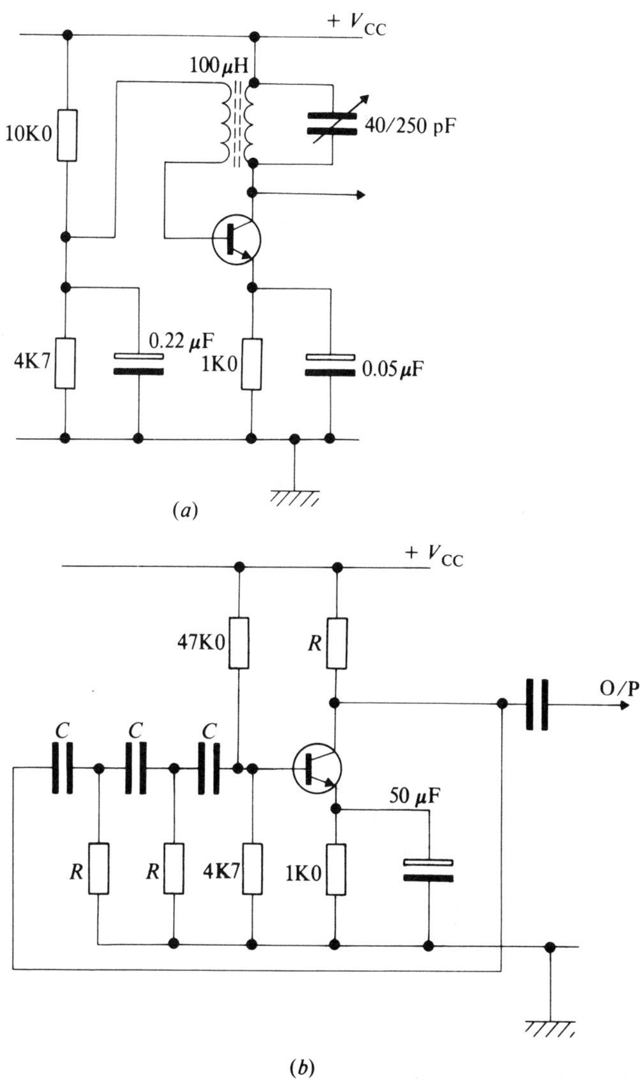

Fig. 9.3 Sine-wave oscillators. (*a*) LC oscillator; (*b*) RC oscillator.

2. ***Square waves*** are produced by *multivibrator oscillators* which use the relaxation principle of the charge and discharge of a CR circuit (see Sec. 9.10).
3. Other waveforms are generally produced from sine or square waves. The most common wave-shaping circuit is the CR (or LR) differentiating and integrating circuit.

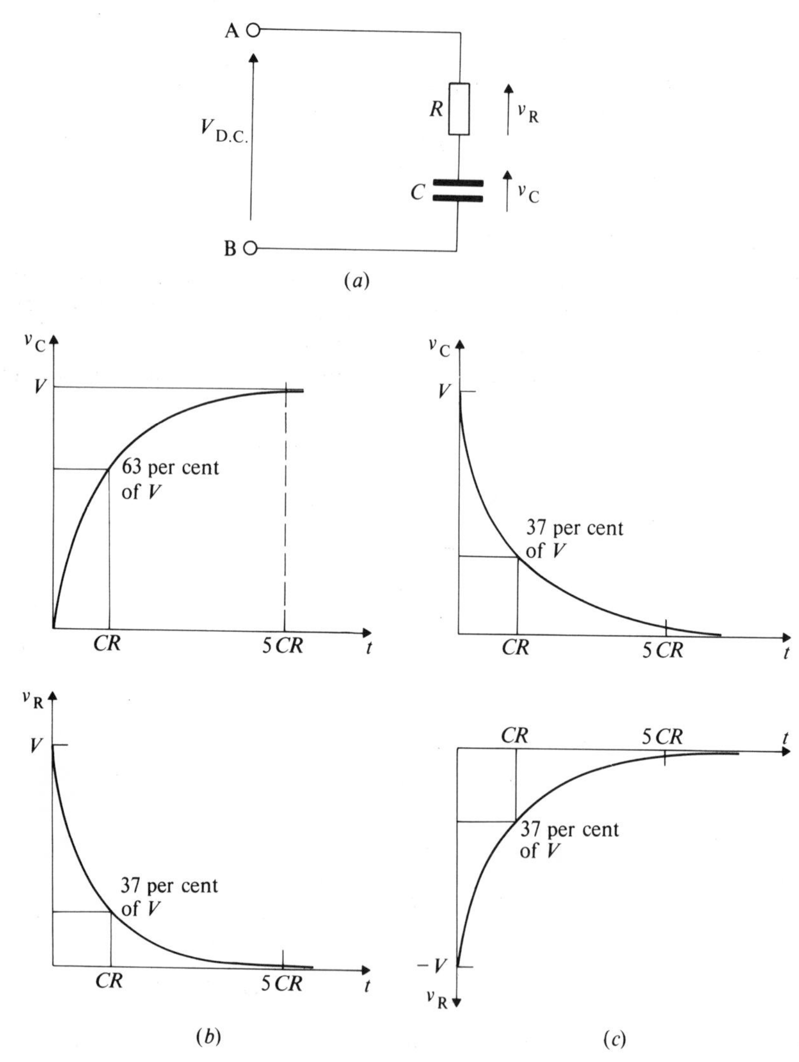

Fig. 9.4 Series CR circuit. (*a*) Circuit; (*b*) charging voltage waveforms; (*c*) discharging voltage waveforms.

9.3 Series CR circuit

We have already seen the charging and discharging waveforms of voltage and current for a series CR circuit in Chapter 3. Consider the simple series CR circuit shown in Fig. 9.4(*a*). If a square wave is applied to the input terminals AB, then the voltage waveforms across C and R could be as shown in Fig. 9.4(*b*).

The time constant of the CR circuit is given by $T = CR$ seconds where C is in farads, R in ohms, and the capacitor is generally considered to be fully charged after 5 CR seconds.

1. A circuit in which the pulse duration is greater than 10 times the time constant is known as a *short CR* circuit, i.e., $t_d > 10\ CR$.
2. A circuit in which the pulse duration is less than 10 times the time constant is known as a *long CR* circuit, i.e., $10CR > t_d$.

9.4 Differentiating and integrating networks

If a square wave is applied to a short CR circuit and the output is taken across R, the waveform is a series of short duration pulses as shown in Fig. 9.5(*c*). This type of circuit is known as a *differentiating network.*

Note: Differentiating is a measure of the *rate of change* of the applied waveform.

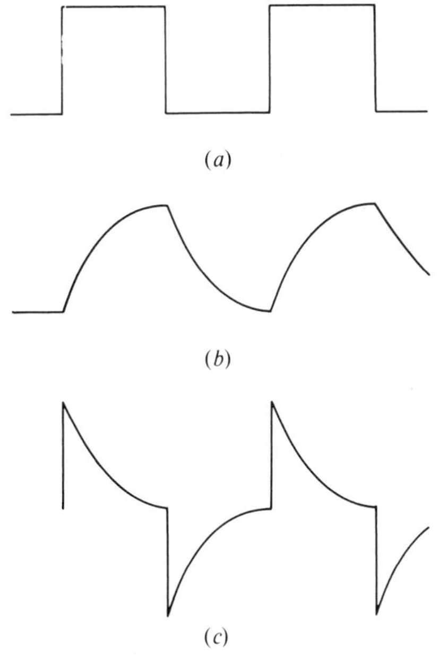

Fig. 9.5 Integrating and differentiating waveforms. (*a*) Input waveform, V_{AB}; (*b*) integrated output waveform, V_C; (*c*) differentiated output waveform, V_R.

Alternatively, if a square wave is applied to a long CR circuit and the output is taken across *C* the waveform is a triangular waveform, as shown in Fig. 9.5(*b*). This type of circuit is known as an *integrating circuit.*

Note: Integrating is a measure of the *area under the applied waveform.*

Practical Exercise 9a

Differentiating and integrating networks

Connect up the circuits shown in Fig. 9.6(*a*) and (*b*).

Apply a square wave at 100 Hz to the input of the circuit shown in Fig. 9.6(*a*), and connect a dual-trace CRO to observe the input waveform, and the differentiated output waveform across the 10K0 resistor. Repeat for sine and triangular input waveforms, record the results, and compare the input and output waveforms.

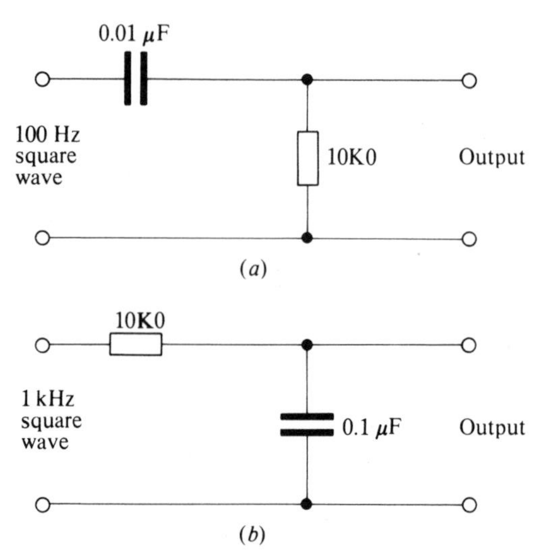

Fig. 9.6 (*a*) Differentiating and (*b*) integrating circuits.

Apply a square wave at 1 kHz to the input of the circuit shown in Fig. 9.6(*b*). Observe the input waveform and the integrated output waveform across the 0.1 μF capacitor. Repeat for sine and triangular input waveforms, record the results, and compare the input and output waveforms.

9.5 Limiting or clipping networks

A limiting circuit is used to cut off part of any voltage waveform, either above or below (or both) any chosen level. Limiting circuits use only a diode and a resistor, the resistance being high compared with the forward resistance of the diode.

These networks may be positive or negative limiters; a *positive limiter* cuts off the waveform from *above* the chosen level, whereas a *negative limiter* cuts off the waveform from *below* the chosen level. In addition, limiting networks may be connected in series or in parallel (shunt). The diode is connected in series between the input and output in the *series limiter* as shown in Fig. 9.7(*a*). The diode is connected in parallel with the output in the *shunt limiter* as shown in Fig. 9.7(*b*) and (*c*). A positive limiter may be easily identified if the positive terminal is at the junction of R and the anode of the diode.

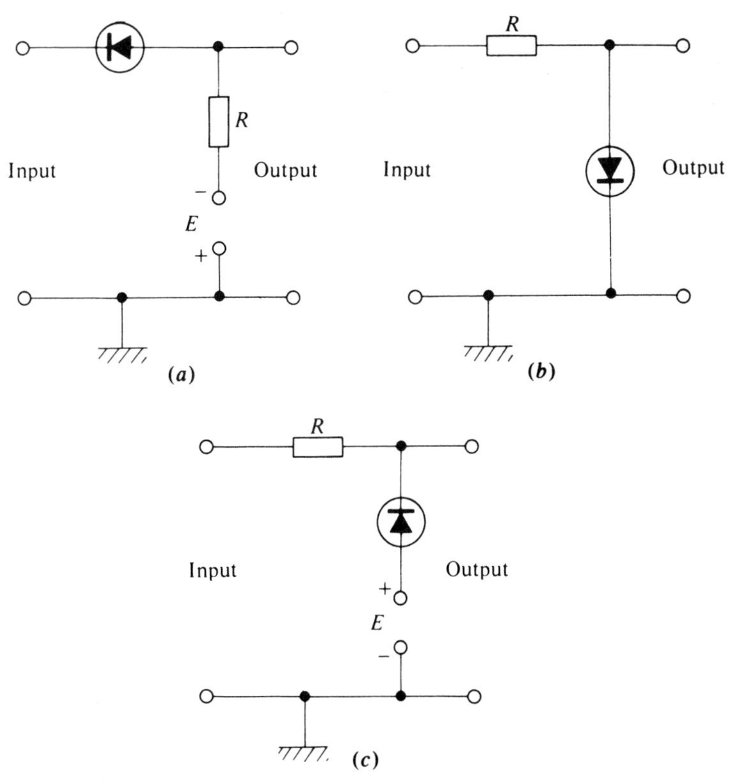

Fig. 9.7 Limiting networks. (*a*) Positive series limiter (limited to $-E$ V); (*b*) positive shunt limiter; (*c*) negative shunt limiter (limited to $+E$ V)

When the limit is a voltage level other than zero, the desired voltage is applied as a bias in the shunt circuit. Positive and negative limiters can be combined, and if the input is sinusoidal, the combined limiter is then known as a diode squarer, as shown in Fig. 9.8.

Limiting networks have a useful application in producing a series of positive (or negative) pulses from a square wave as shown in Fig. 9.9, which is a differentiating network followed by a negative limiter.

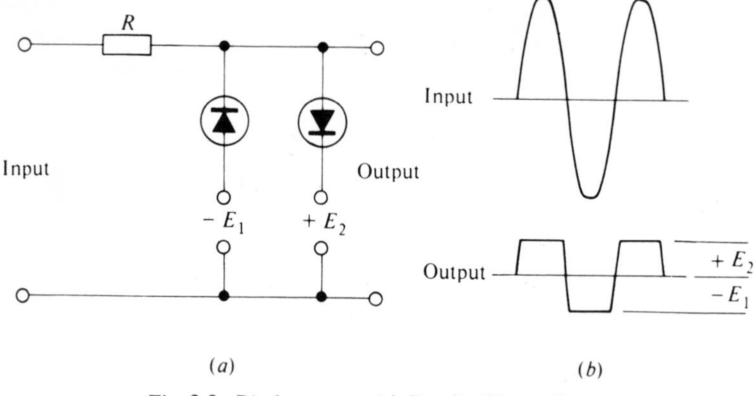

Fig. 9.8 Diode squarer. (*a*) Circuit; (*b*) waveforms.

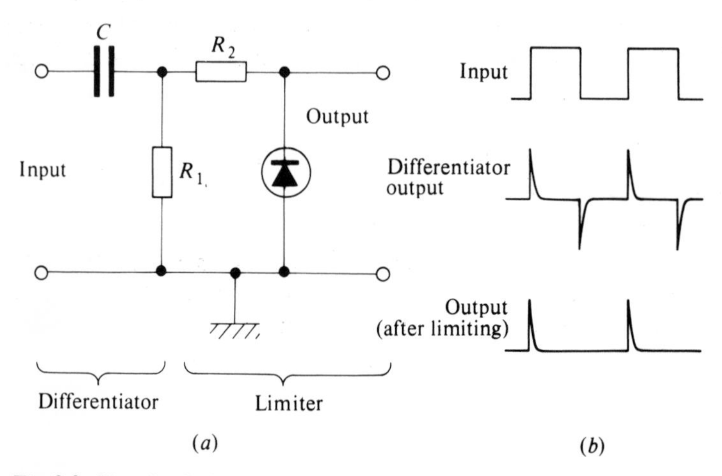

Fig. 9.9 Negative limiter combined with differentiator. (*a*) Circuit (*b*) waveform.

Practical Exercise 9b

Pulse waveform generator

Connect up the circuit shown in Fig. 9.10.

Apply a square wave of frequency 100 Hz to the input terminals. Using a dual-trace CRO, observe the input and output waveforms. Record the results.

Reverse the diode and repeat, noting the effects of the reversal. Replace the 10K0 resistor with a 1K0 resistor and repeat.

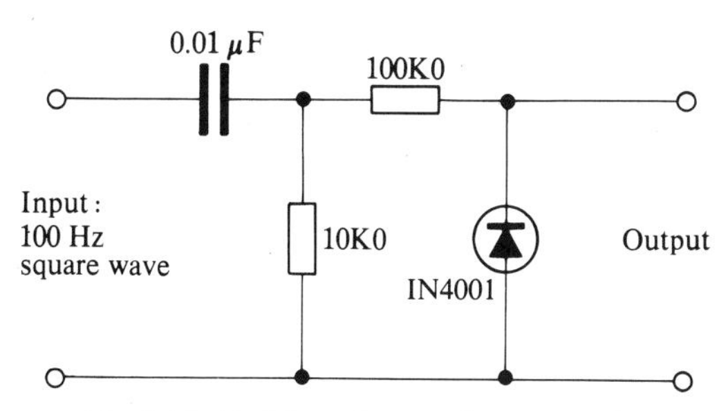

Fig. 9.10 Positive pulse waveform generator.

9.6 Squaring circuits

As the name suggests, this type of circuit is used to produce square waves from sine-wave inputs. Two main types are widely used. The diode and transistor squarers effectively cut off the top and bottom of the sine-wave input. The second type of squarer is the *Schmitt trigger* circuit which is capable of producing a good square (or rectangular) waveform from any slowly changing input signal.

Practical Exercise 9c

The transistor squarer

Connect up the circuit shown in Fig. 9.11.

Apply a sine-wave input at a frequency of 1 kHz with an amplitude of 10 V, and use the dual-trace CRO to observe the input and output waveforms.

Note the effect of varying the 1M0 variable bias resistor.

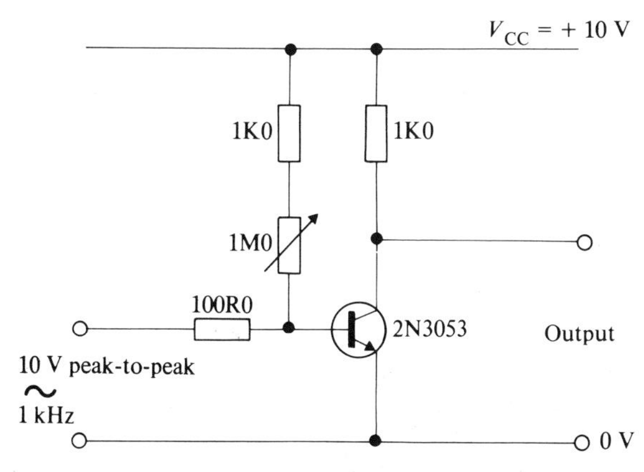

Fig. 9.11 Transistor squarer.

1. Positive half-cycles of the output waveform are limited by cut-off.
2. Negative half-cycles of the output waveform are limited by saturation.
3. The output waveform symmetry may thus be controlled by varying the bias applied to the transistor.

9.7 Clamping networks

Clamping networks are used to move a voltage waveform, up or down, and *clamp* it at any desired level, as shown in Fig. 9.12.

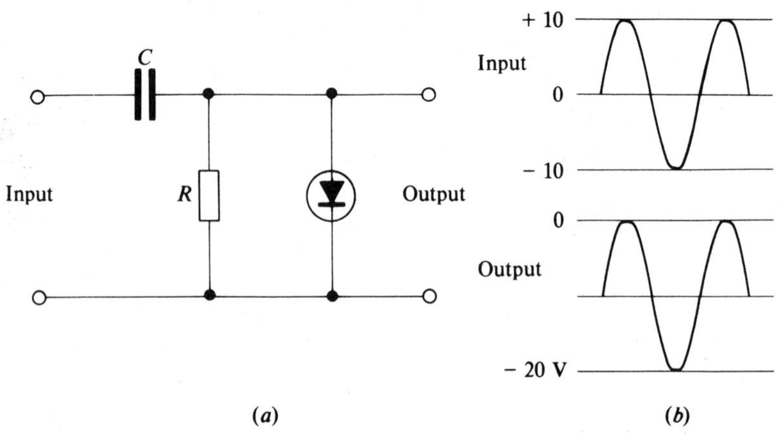

Fig. 9.12 Clamping network. (*a*) Circuit; (*b*) waveforms.

9.8 Pulse-generating circuits

Pulse-generating circuits play an important part in modern electronics. Feedback signals in many control systems are pulse coded; trigger pulses are used to switch motor control thyristors; pulse techniques are used in the ever increasing field of digital logic circuits. Pulse-coded signals are now being used in radio and television communication systems.

The main types of pulse generating circuits depend on the principle of cross-coupled transistors.

9.9 Review of transistor switching

Consider the simple transistor circuit shown in Fig. 9.13(*a*). In a simple transistor switching circuit, the value of the d.c. voltage at the collector is determined from

$$V_{CE} = V_{CC} - I_C R_L.$$

However, if another stage is coupled to the transistor collector, as shown

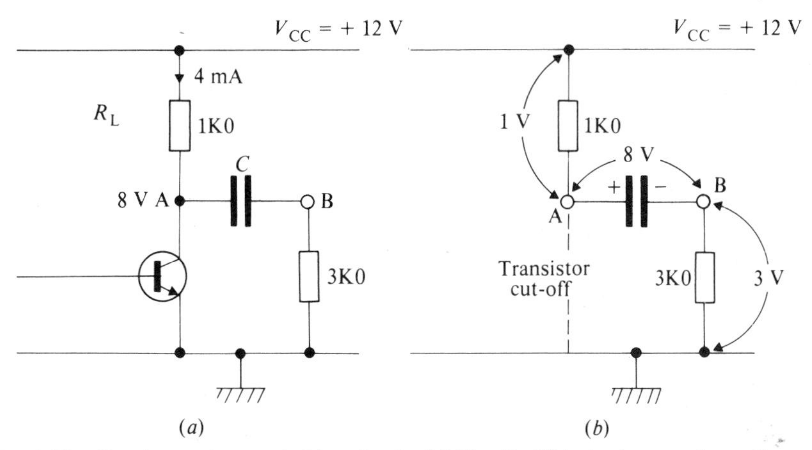

Fig. 9.13 Simple transistor switching circuit. (*a*) Circuit; (*b*) instantaneous d.c. voltages as transistor is switched off.

in Fig. 9.13(*a*), the charge and discharge of the coupling capacitor connected to the collector will cause the value of V_{CE} to differ from its expected value.

When the transistor is cut off, its collector voltage immediately tries to rise to V_{CC}. This will occur if the transistor has a resistive load only. But, if a capacitor is connected to the collector, V_{CE} is only able to rise as quickly as the time constant of the circuit will allow.

EXAMPLE

A transistor is conducting a steady current of 4 mA. Under these conditions the voltage at the collector is 8 V (with a 1K0 load resistor) and the coupling capacitor *C* is charged to 8 V (through the 3K0 resistor) under steady conditions.

If the transistor is now suddenly cut off, the circuit effectively becomes equivalent to that shown in Fig. 9.13(*b*). Therefore, the collector voltage is no longer at 8 V, since the remaining 4 V ($V_{CC} - V_{CE}$ = 12 – 8 = 4 V) is immediately shared between the two series connected resistors, and the voltages at each point in the circuit are as shown in Fig. 9.13(*b*).

The voltage at the collector, therefore, immediately rises to + 11 V (not to V_{CC} = + 12 V) and the voltage at the junction of the capacitor and the 3K0 resistor immediately rises to + 3 V. The capacitor *C* now charges in the usual way through the two series resistors, and after 5 *CR* seconds the voltage across the capacitor will be + 12 V; thus, the collector voltage will also be + 12 V. (The junction of *C* and the 3K0 resistor will now be 0 V.)

9.10 Multivibrators

These are a class of electronic switching circuits which are also known as relaxation oscillators, since the transistors are driven beyond cut-off for a period of time during their operation. The output waveform is generally square, or rectangular pulses.

Multivibrators use two transistors (or valves) and at any time during their operation one transistor is saturated and the other is cut off. There is a family of multivibrator circuits and we shall briefly examine each of these:

1. ***The astable multivibrator.*** This is often called a *free running* multivibrator since it requires no separate input signal, and produces a continuous train of square waves at its output.
2. ***The bistable multivibrator.*** This multivibrator is often called a *flip-flop*, since it has two stable operating states. When an input signal is applied, the output changes from one stable operating state to the other.
3. ***The monostable multivibrator.*** This is sometimes called a *one-shot* multivibrator, in which the application of an input signal causes the output to change to a virtually stable state in which it remains for a period of time (dependent on component values). After this time, the output returns to its original stable state.

9.11 The astable multivibrator

The circuit of an astable multivibrator is shown in Fig. 9.14, and consists of two transistor amplifiers connected in cascade, in which all of the output of one stage is fed back (100 per cent positive feedback) to the input of the other stage.

When the supply is switched on, unbalance between the two stages causes one transistor to conduct whilst the other remains cut-off. Suppose that TR 1 saturates, then the fall in its collector voltage is transferred to the base of TR 2 which ensures that TR 2 is cut-off. During the time that the circuit remains in this state, capacitor C_1 charges from $-V_{CC}$ *towards* $+V_{CC}$ at a rate depending on time constant C_1R_3. The time taken for the capacitor voltage to reach about

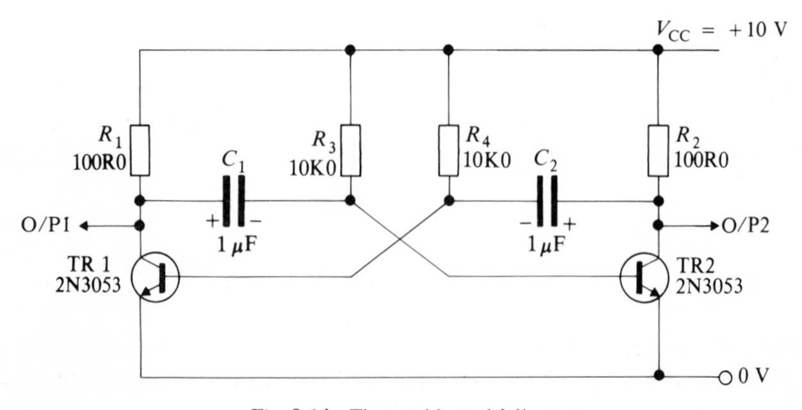

Fig. 9.14 The astable multivibrator.

0.7 V is given by

$$T_1 \approx 0.7C_1R_3 \text{ seconds.}$$

After this time, the capacitor voltage causes TR 2 to start conducting, and the regenerative action causes TR 1 to be cut off, so that the collector voltage rises towards + V_{CC} at a rate depending on time constant C_1R_1. The time that the circuit remains in this state is given by

$$T_2 \approx 0.7C_2R_4 \text{ seconds.}$$

This sequence is repeated, and the waveforms for the circuit are as shown in Fig. 9.15.

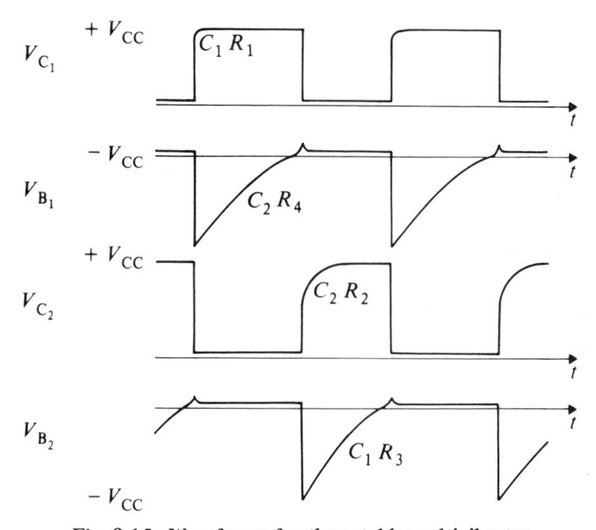

Fig. 9.15 Waveforms for the astable multivibrator.

If $C_1 = C_2$ and $R_3 = R_4$, then the duration of each state is the same and $T_1 = T_2$. Therefore, the periodic time for the output is $2T_1$ and the frequency is given by:

$$f = \frac{1}{2T_1} = \frac{1}{1.4C_1R_3} \text{ Hz.}$$

If the output is taken from TR 1 collector, then the ratio T_2/T_1 is the mark-to-space ratio of the waveform. The mark-to-space ratio may be changed without significantly changing the frequency by using the modification shown in Fig. 9.16(*a*). The frequency may be changed without significantly changing the mark-to-space ratio by using the modification shown in Fig. 9.16(*b*).

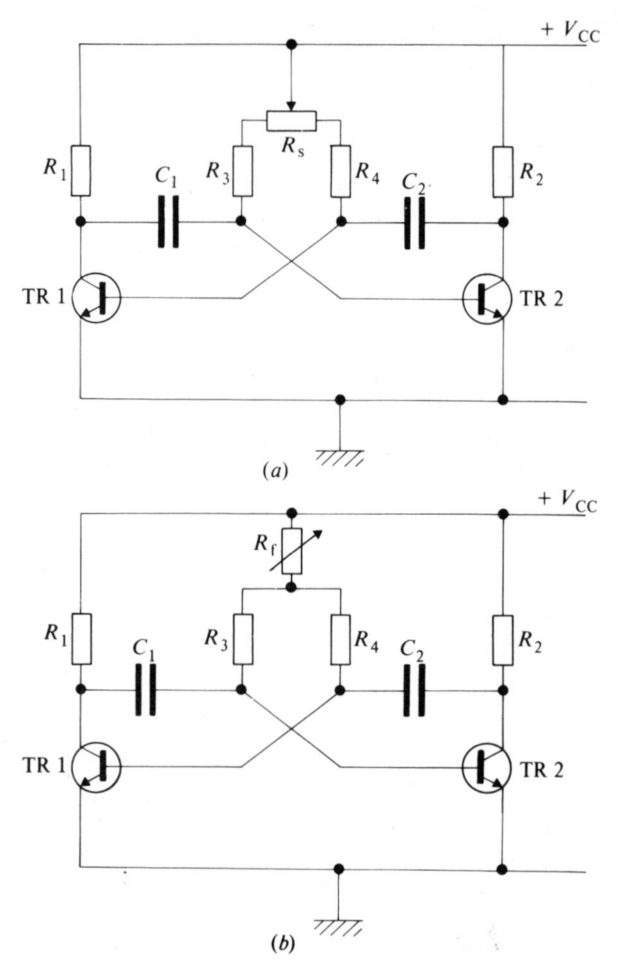

Fig. 9.16 Modifications to the astable multivibrator. (*a*) Mark-to-space variation; (*b*) frequency variation.

Practical Exercise 9d

The astable multivibrator

Connect up the circuit shown in Fig. 9.14.

Switch on the supply and use a dual-trace CRO to examine the waveforms at the base and collector of each transistor. Record these waveforms and compare with those shown in Fig. 9.15.

Replace R_3 by a 50K0 variable resistance and note the effect of varying this resistance on the waveforms. Does the mark-to-space ratio change?

Repeat the above, but with the modifications shown in Fig. 9.16 using the 50K0 variable resistance.

Replace C_1 and C_2 by 100 μF electrolytic capacitors and R_1 and R_2 by 12 V lamps to set up a flashing lamp display.

9.12 The bistable multivibrator

The circuit of a bistable multivibrator is shown in Fig. 9.17.

When the supply is switched on, only one transistor conducts due to the imbalance between the two stages. Assume that TR 2 is conducting, then its low collector voltage ensures that TR 1 is cut off. This is one stable state, with output 1 high (near to + V_{CC}) and output 2 low (near to 0 V).

A positive pulse applied to the SET terminal will cause TR 1 to begin conducting, its collector voltage falls and is applied to TR 2 base, causing TR 2 to be cut off, and its collector (output 2) thus goes high. This is the second stable operating state. A positive pulse applied to the RESET terminal will cause the circuit to revert to its original operating state – hence the term flip-flop (in this case a *S–R flip-flop*).

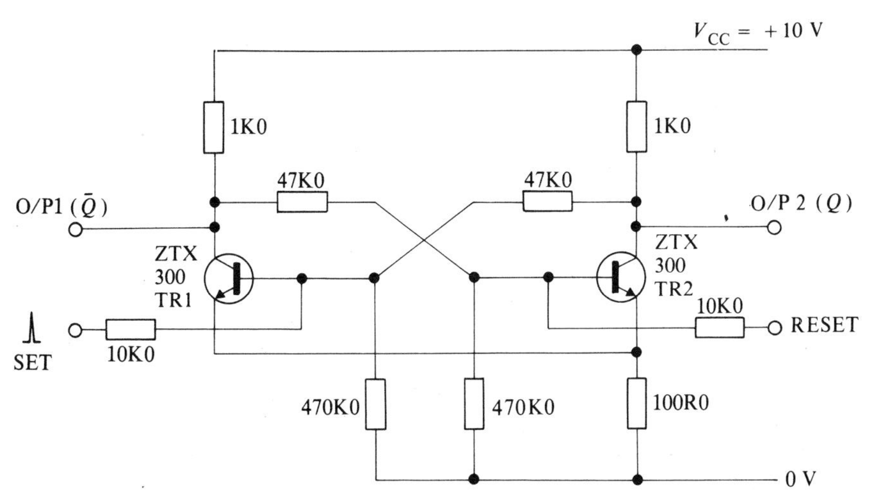

Fig. 9.17 The bistable multivibrator.

Practical Exercise 9e

The bistable multivibrator

Connect up the circuit shown in Fig. 9.17.

Switch on the supply and measure the collector voltages of both transistors.

Apply a positive pulse voltage to the SET terminal and check that the collector voltage of TR 2 goes high. Now apply the positive pulse to the RESET terminal and check that the collector voltage of TR 2 goes low (and TR 1 collector voltage goes high).

9.13 The monostable multivibrator

The circuit of a monostable multivibrator is shown in Fig. 9.18. This type of circuit is used for pulse shaping, or to provide a specified time delay which is independent of the input trigger pulse.

The stable operating state of this circuit is achieved with TR 2 saturated (and TR 1 cut off). TR 2 is saturated by the base drive through R_3, and the collector voltage of TR 2 is low. The negative bias applied to TR 1 base ensures that TR 1 remains cut off. Capacitor C charges up to $+V_{CC}$ volts with the polarity shown.

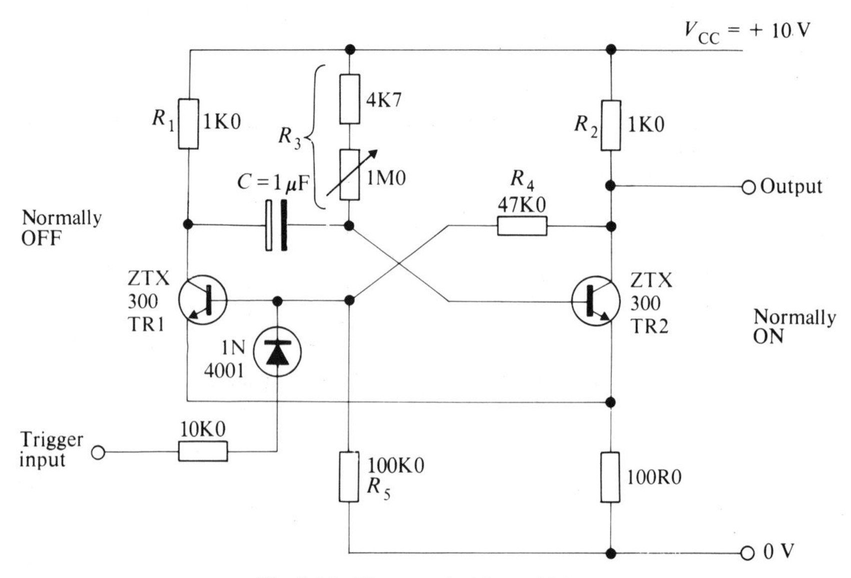

Fig. 9.18 The monostable multivibrator.

The circuit may be triggered into its virtually stable state by the application of a positive pulse to TR 1 base, which drives TR 1 into saturation and this connects the left hand plate of capacitor C to earth. This action applies the capacitor voltage ($-V_{CC}$ volts) to TR 2 base, thus cutting off TR2. At the same time the increased collector voltage of TR 2 is applied to TR 1, which keeps TR 1 in saturation.

The capacitor voltage (at TR 2 base) now charges from $-V_{CC}$ *towards* $+V_{CC}$ at a rate dependent on the time constant CR_3. When the capacitor voltage reaches about 0.7 V, TR 2 will be switched on and TR 1 is cut off, so that the circuit reverts to its stable operating state. The waveforms for this circuit are as shown in Fig. 9.19.

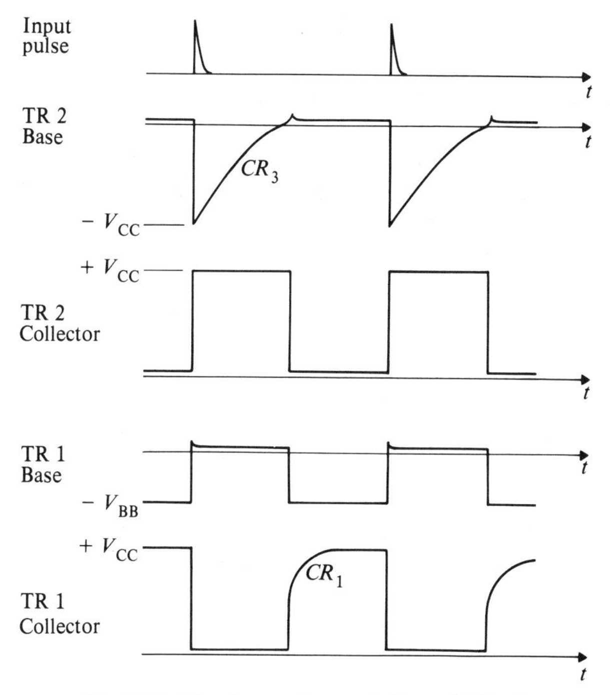

Fig. 9.19 Waveforms of monostable multivibrator.

Practical Exercise 9f

The monostable multivibrator

Connect up the circuit shown in Fig. 9.18.

Switch on the supply, and measure the collector voltage of TR 2.

Connect the collector and base voltages of TR 2 to a dual-trace CRO, apply a positive pulse to the input trigger, and observe the effect on the waveforms. Compare the waveforms with those shown in Fig. 9.19. Repeat this procedure while monitoring the collector and base voltages of TR 1.

9.14 The Schmitt trigger

The circuit of a Schmitt trigger is shown in Fig. 9.20(*a*), and produces square waves from any shape of waveform applied to the input.

The normal state (in the absence of an input signal) of this circuit is that TR 2 is biased so that it is conducting – but not saturated. When the input waveform rises above a predetermined value, the output voltage suddenly rises to almost $+V_{CC}$. The circuit reverts to its normal state when the input voltage waveform falls below another predetermined level. The difference between the

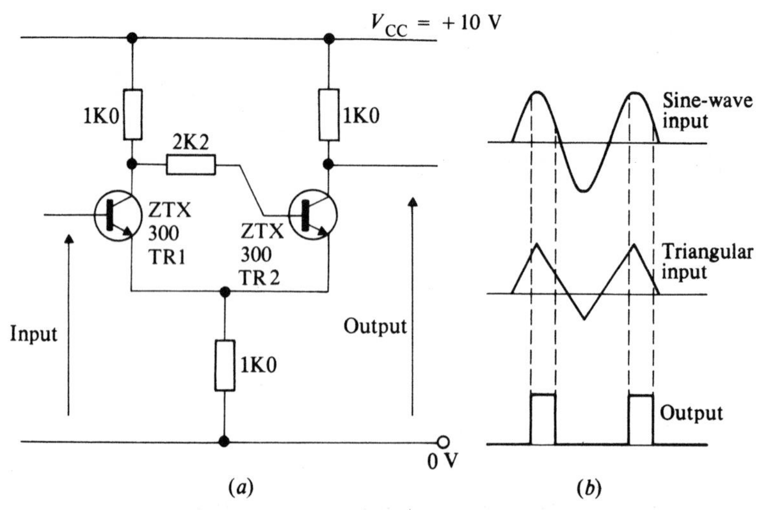

Fig. 9.20 The Schmitt trigger. (*a*) **Circuit;** (*b*) **waveforms.**

ON and OFF levels is termed *hysteresis* or *backlash*. Typical waveforms are shown in Fig. 9.20(*b*).

Many of the circuits which we have considered in this chapter are available in integrated circuit form, i.e., instead of all the discrete components shown the circuits are produced in a 'chip' of silicon and encapsulated in a single package.

It is possible to use the popular 741 operational amplifier as a Schmitt trigger.

Practical Exercise 9g

The Schmitt trigger

Connect up the circuit shown in Fig. 9.20(*a*).

Apply a sine-wave signal at 1 kHz to the input, and observe the input and output waveforms using a dual-trace CRO. Increase the amplitude of the signal until the circuit triggers. Record the waveforms and compare with those shown in Fig. 9.20(*b*).

Repeat with a triangular input waveform.

Practical exercise 9h

The 741 operational amplifier as a Schmitt trigger

Connect up the circuit shown in Fig. 9.21.

Apply a sine-wave signal at 1 kHz to the input, and observe the input and output waveforms using a dual-trace CRO. Increase the amplitude of the input signal until the circuit triggers. Record the waveforms and compare with those given in Fig. 9.20(*b*).

Repeat with a triangular input waveform.

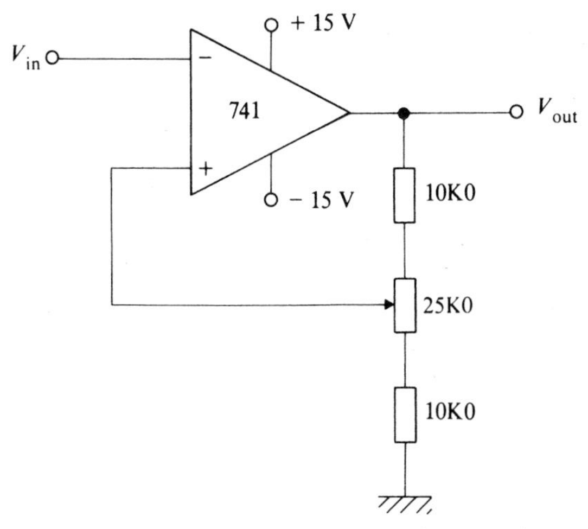

Fig. 9.21 741 operational amplifier as a Schmitt trigger.

10. Amplifiers

10.1 Introduction

Many types of amplifiers will be encountered in the field of industrial electronics. Amplifiers may be classified according to: frequency range, methods of inter-stage coupling used, bias point at which the active device operates, voltage, current and power. Circuits which amplify a wide range of frequencies are known as *wideband* (or *untuned*) amplifiers, and those which are 'tuned' to amplify a narrow band of frequencies are known as *narrowband* (or *tuned*) amplifiers. Methods of coupling amplifier stages modify the performance to some extent. The most common method is *a.c. (alternating current) coupling*, in which low frequency components (including d.c.) are not transmitted to the following stage. Some amplifiers are *d.c.* (*direct current*) *coupled*, in which all frequency components down to d.c. are transmitted to the following stage. One form of d.c. amplifier is the *chopper* amplifier, in which the input signal is 'chopped' into a series of pulses, i.e., an alternating signal, which is amplified by a.c. coupled amplifiers before being re-converted back into d.c.

The operating conditions to which the active devices are biased is related to the amplifying function carried out by that stage:

- ***Class A*** – current flows in the load during the whole period of the input signal cycle.
- ***Class AB*** – current flows in the load for more than half a cycle, but less than the full cycle of the input signal.
- ***Class B*** – current flows in the load for half a cycle of the input signal.
- ***Class C*** – current flows in the load for less than half a cycle of the input signal.

Tuned and untuned voltage amplifiers, and low power a.f. (audio frequency) amplifiers generally work in Class A, whilst a.f. power amplifiers work in Class B. Oscillators and r.f. (radio frequency) amplifiers usually operate in Class C.

The *gain* of an amplifier is the ratio of the magnitude of the output signal to that of the input signal.

Voltage amplifiers increase the magnitude of the input signal voltage, but are generally not capable of providing a significant amount of output power. *Current* and voltage amplifiers are generally grouped together – the main difference being the relative values of their output impedance. *Power* amplifiers are designed to provide an adequate signal power to drive output devices.

Operating efficiency of power amplifiers is a main consideration, whereas distortion may not be important. The opposite is generally the case for voltage amplifiers.

In Chapter 6, we saw that a transistor can be used to amplify an input signal. A *standard* common emitter voltage amplifier circuit is shown in Fig. 10.1, in which R_L is the load resistance across which voltage changes are produced from collector current changes. R_1 and R_2 together with R_E set the d.c. bias conditions to determine the transistor operating point. R_E is also used to provide temperature stabilization, and the variations of emitter current at the signal frequency are decoupled by C_E.

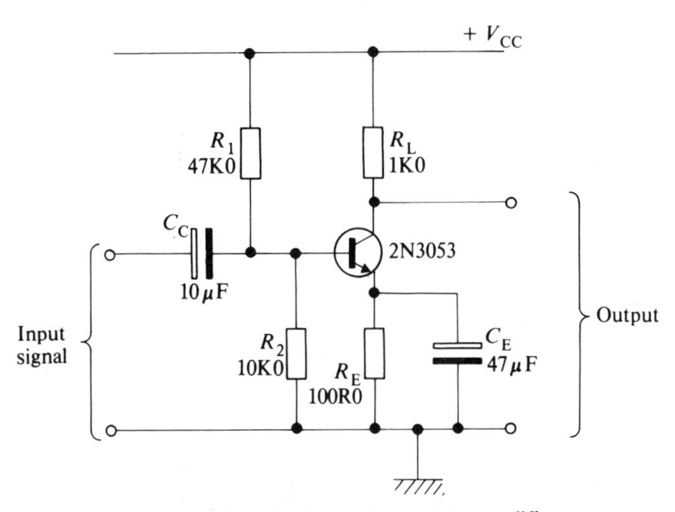

Fig. 10.1 Standard common emitter amplifier.

10.2 Matching

The amplifier arrangement shown in Fig. 10.1 is perfectly adequate for many applications where it is required to amplify a low-level signal voltage. However, in many applications, e.g., control engineering, additional problems are created with *signal conditioning*, as it is called, because of the impedances of the input and output circuits as follows:

1. matching a low-level voltage–high-impedance source to a pre-amplifier;
2. matching a low-impedance load, e.g., relay or loudspeaker, to an amplifier to produce maximum power in the load.

10.3 Matching source and input impedances

As stated above, signal conditioning causes problems in many control engineering

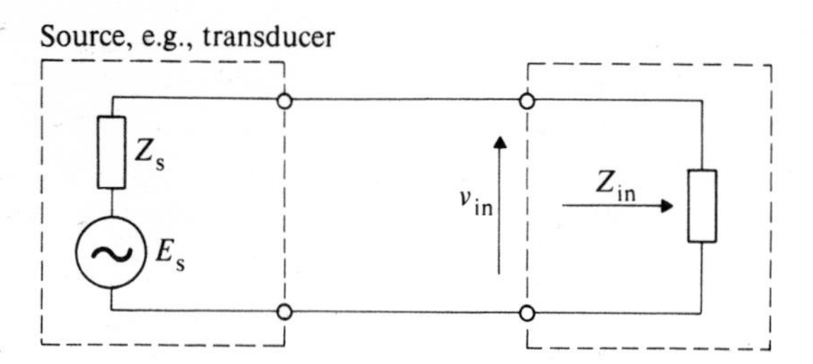

Fig. 10.2 Source and input impedances.

situations. Typical of this type of application is feeding the small output from a transducer to an amplifier, as shown in Fig. 10.2.

Assume that the e.m.f. produced by the transducer $E_s = 1$ V, and that the transducer impedance $Z_s = 1$ MΩ.

If we use a standard common emitter voltage amplifier, the input impedance Z_{in} will be about 1 kΩ, and the voltage v_{in} at the amplifier input is given by

$$v_{in} = \frac{E_s \times Z_{in}}{Z_s + Z_{in}} = \frac{1 \times 1000}{1\,001\,000}\ \text{V} \approx 1\ \text{mV}.$$

Now, if we could use an amplifier which has an input impedance of, say, 100 kΩ, then the input voltage v_{in} is given by:

$$v_{in} = \frac{E_s \times Z_{in}}{Z_s + Z_{in}} = \frac{1 \times 100\,000}{1\,100\,000}\ \text{V} \approx 100\ \text{mV}.$$

which is a considerable improvement.

Therefore, when a pre-amplifier is used to amplify a low-level signal from a high impedance source, the pre-amplifier needs a ***high input impedance*** so as not to *load* or *damp* the signal from the source.

Practical Exercise 10a

Impedance matching

Connect up the circuit shown in Fig. 10.3.

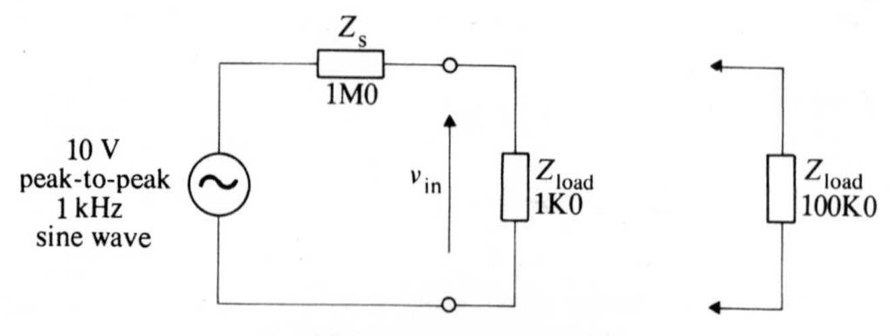

Fig. 10.3 Impedance matching.

Set the function generator to SINE, 1kHz, 10 V peak-to-peak and connect it to the input terminals. Using a dual-trace CRO, measure the voltage applied to the circuit, and the voltage developed across the 1KO load resistor.

Replace the 1KO resistor by a 100KO resistor and repeat the measurements. What effect does the value of load resistance have on the voltage v_{in}?

10.4 The emitter follower

This is a transistor amplifier in which the load is connected in the emitter circuit, as shown in Fig. 10.4. It is sometimes referred to as a common collector amplifier. This method of connection gives a high input impedance, typically 100 kΩ, and is often used as a *buffer* amplifier, since it has an additional advantage of being capable of driving low impedance loads such as relays or loudspeakers.

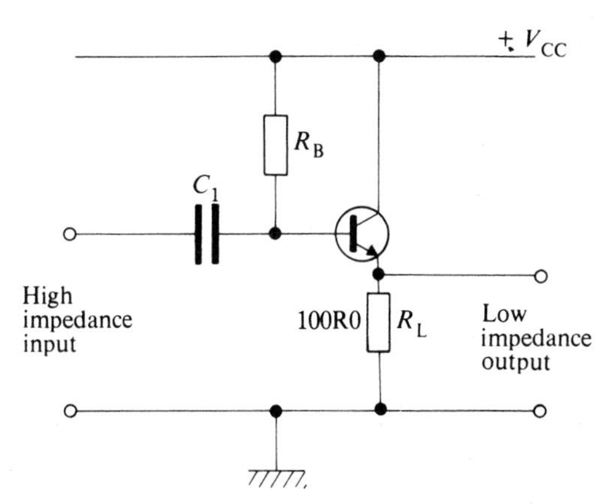

Fig. 10.4 Emitter follower circuit.

The three basic transistor amplifier configurations are shown in Fig. 10.5, and a comparison of their characteristics is given below:

1. ***Common base.***

   Current gain, $h_{FB} \approx 0.99$.
   Voltage gain = 50.
   Input impedance, Z_{in} = 50 Ω.
   Output impedance, Z_{out} = 250 kΩ.
   Power gain ≈ 50.

2. ***Common emitter.***

Current gain, $h_{FE} \approx 200$.
Voltage gain = 50.
Input impedance, Z_{in} = 1 kΩ.
Output impedance, Z_{out} = 50 kΩ.
Power gain ≈ 2500.

3. ***Common collector (emitter follower).***

Current gain, $h_{FE} \approx 200$.
Voltage gain = 1.
Input impedance, Z_{in} = 100 kΩ.
Output impedance, Z_{out} = 1 kΩ.
Power gain ≈ 50.

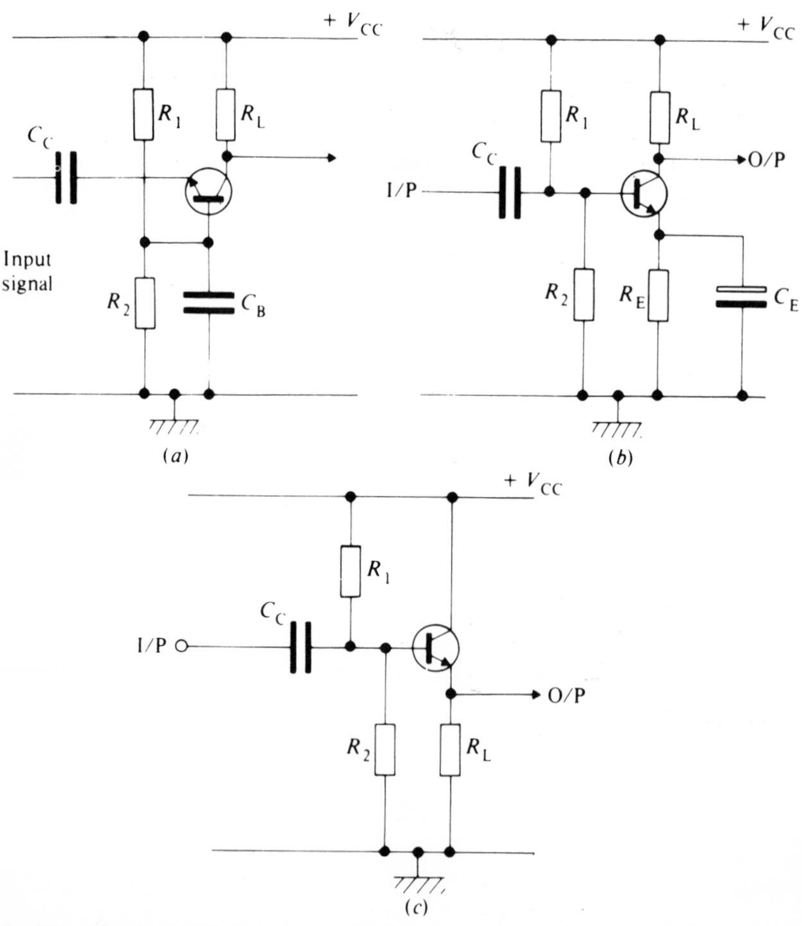

Fig. 10.5 The three basic transistor amplifier configurations. (*a*) Common base; (*b*) common emitter; (*c*) common collector.

Practical Exercise 10b

The emitter follower

Connect up the circuit of the signal source stage with the common emitter amplifier stage shown in Fig. 10.6(*a*).

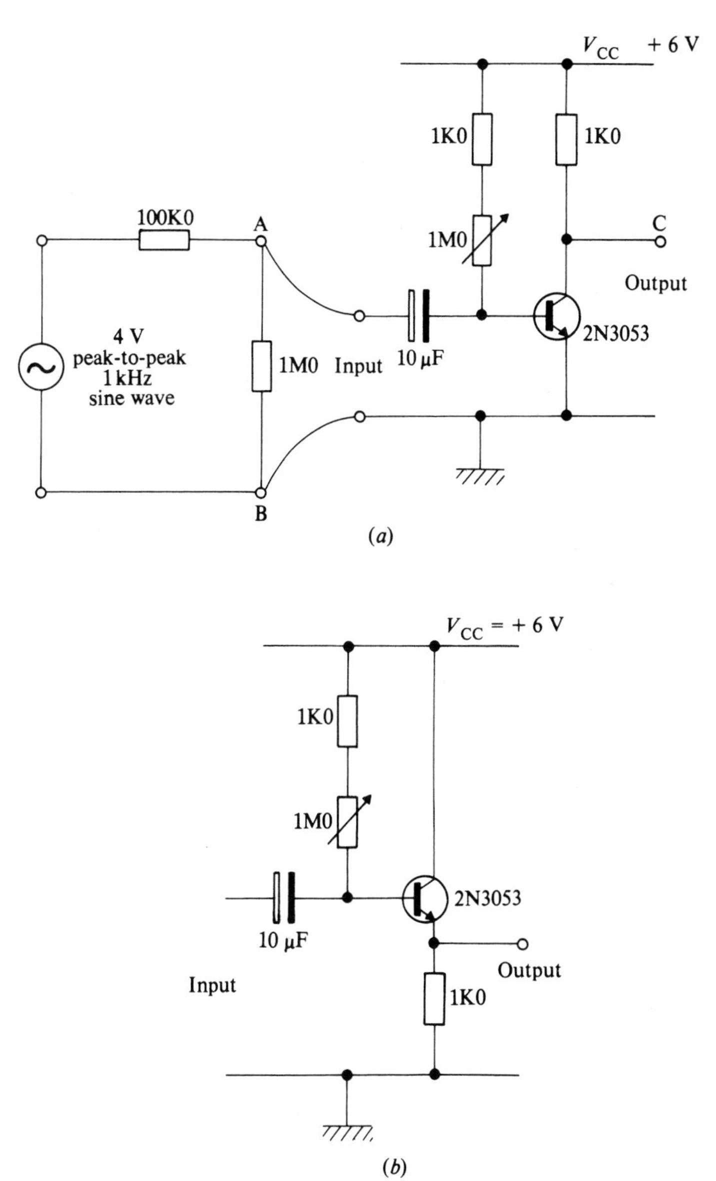

Fig. 10.6 Effect of amplifier input impedance. (*a*) Signal source feeding a common emitter amplifier; (*b*) emitter follower.

Set the function generator to SINE, 1 kHz, 4 V peak-to-peak and connect to the signal input terminals. Using a dual-trace CRO, measure the input signal across AB, and the output signal across CB. Adjust the 1M0 variable resistor (bias) to give an undistorted output waveform.

Record the amplitude of the input signal (AB) and the output signal (CB). Disconnect the lead from the transistor base and observe the effect on the amplitude of the input signal.

Rearrange the amplifier stage to that of the emitter follower shown in Fig. 10.6(*b*) and repeat the above measurements.

What are the differences between the two amplifier stages and what conclusions about input impedance can be drawn from the measurements made above?

10.5 The Darlington pair

This form of connection, formerly known as the 'super alpha' pair, is shown in Fig. 10.7. The Darlington pair circuit may use discrete transistors, or the transistors may be combined in one single integrated circuit (IC) chip.

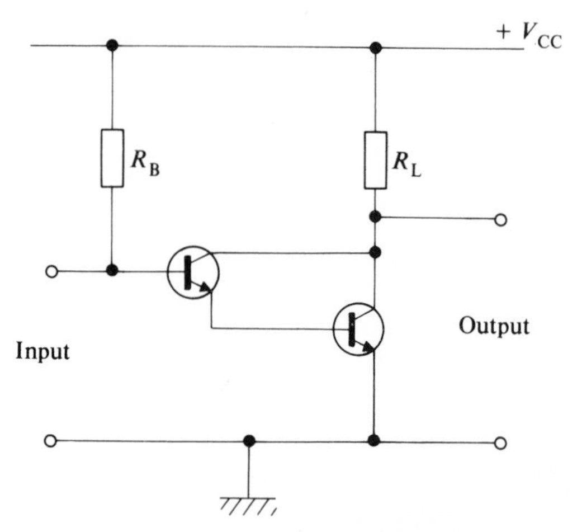

Fig. 10.7 The Darlington pair.

This method of connection gives a very high input impedance, typically 1 MΩ, and produces very high current gain, typically several thousand. The current gain is given approximately by $h_{FE1} \times h_{FE2}$.

10.6 Feedback in amplifiers

Part of the output is connected to the input as *feedback* in amplifiers, to vary the performance. If the feedback is such that it opposes the original input signal, it is *negative feedback* which generally reduces the gain, but improves the stability of gain. If the feedback is such that it assists the original signal, it is *positive feedback* which is generally used to produce oscillators.

Consider the simple amplifier shown in Fig. 10.8(*a*). The ratio of output voltage to input voltage is the *gain* of the amplifier. This arrangement will be referred to as an open-loop amplifier, and

$$\text{open-loop gain, } A = \frac{e_o}{e_i}$$

If part of the output is now fed back through a feedback network β, such that when it is combined (mixed, or compared) with the original input, it opposes the input, then the arrangement shown in Fig. 10.8(*b*) is produced. This is called a closed-loop amplifier, in which the overall gain is $v_o v_i$ i.e. v_o/v_i.

Amplifier input, $v = v_i - \beta v_o$

and Amplifier output, $v_o = A\ v$

so $$v_o = A\ (v_i - \beta v_o)$$
$$= Av_i - \beta A v_o$$

Thus $$v_o + \beta A v_o = A v_i$$

and $$(1 + \beta A) v_o = A v_i$$

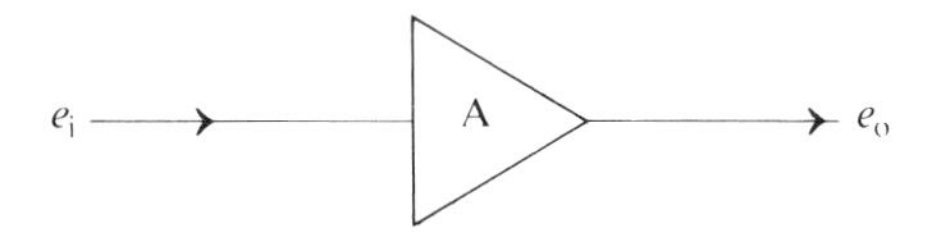

(*a*) Open-Loop amplifier

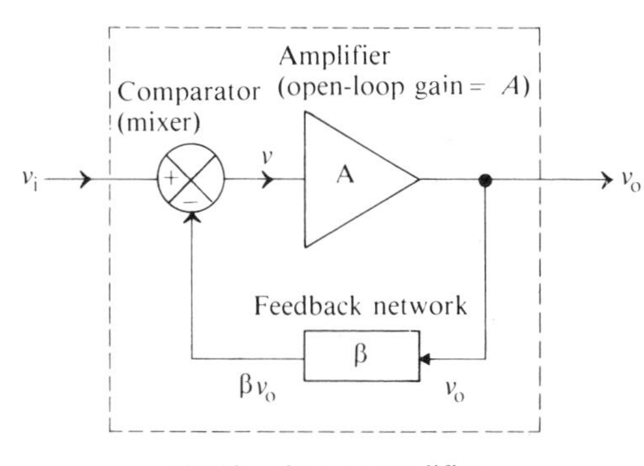

(*b*) Closed-Loop amplifier

Fig. 10.8 Concepts of feedback in amplifiers.

Therefore: gain (with negative feedback) $= \dfrac{v_o}{v_i} = \dfrac{A}{1 + \beta A}$ (10.1)

The expression (10.1) clearly shows that the gain is reduced (since A is divided by a quantity greater than 1). Gain of amplifiers can be affected by many factors: supply voltage variations, device parameter variations, etc. Improvement in stability of gain may be illustrated by the simple Example 10.1.

EXAMPLE 10.1

Consider an amplifier having an open-loop gain of 100; owing to several factors the gain varies by ± 10 per cent.

Calculate the gain when the amplifier uses 15 per cent *negative* feedback.

Also, calculate the stability of gain for the closed-loop amplifier.

Solution

$$A = 100 \pm 10 \text{ per cent}$$
$$\beta = 15 \text{ per cent}$$

Nominal closed-loop gain $= \dfrac{A}{1 + \beta A} = \dfrac{100}{1 + 15} = \dfrac{100}{16} = 6.25$

Minimum closed-loop gain $= \dfrac{A}{1 + \beta A} = \dfrac{90}{1 + 13.5} = \dfrac{90}{14.5} = 6.21$

Maximum closed-loop gain $= \dfrac{A}{1 + \beta A} = \dfrac{110}{1 + 16.5} = \dfrac{110}{17.5} = 6.29$

Variation in closed-loop gain = 0.04

Variation in gain $= \dfrac{0.04}{6.25} \times 100 = 0.64$ per cent

Therefore, the gain (with negative feedback) is reduced from 100 to 6.25, but the effect of the variation in gain is improved from ± 10 per cent to less than ± 1 per cent, i.e., producing a more stable value of gain. Further stages of amplification can now be used to compensate for the reduction of gain caused by the use of negative feedback.

If the product βA in expression (10.1) becomes negative, then the feedback is *positive*. When $\beta A = -1$, the amplifier will have an infinite gain – suggesting that an output is produced with no input. Positive feedback is used to produce oscillators.

Several possibilities exist for the application of feedback, as shown in Fig. 10.9. A *voltage* or *current* may be derived from the output and fed back to the input as a *series* or *shunt* signal. Positive feedback increases the gain, negative feedback decreases the gain. The voltage feedback circuits shown in Fig. 10.9(*a*) and (*b*) will cause the output impedance to *decrease* – since they have a shunt connection

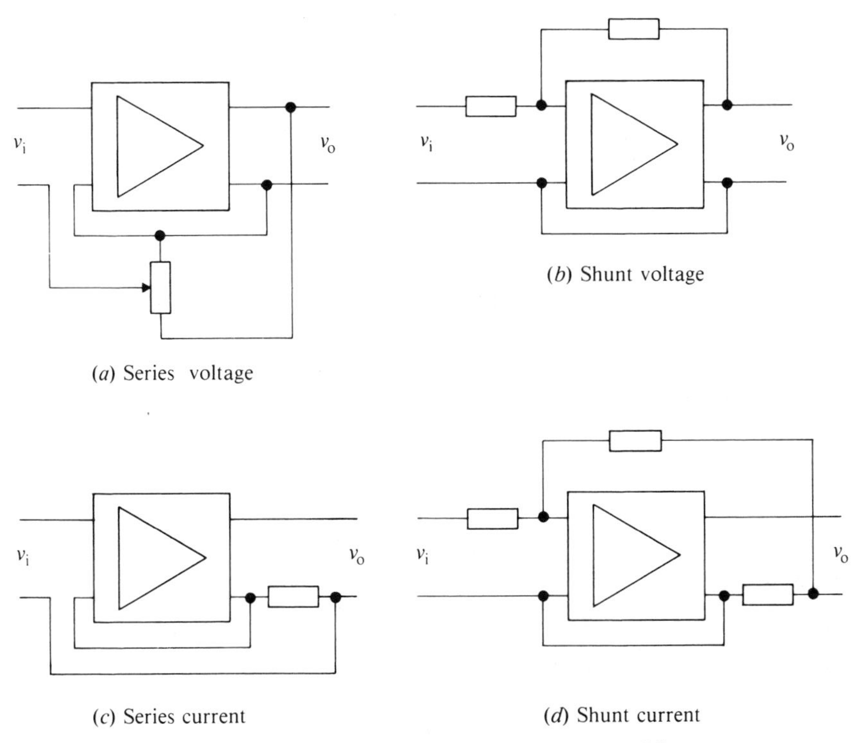

Fig. 10.9 Methods of applying feedback in amplifiers.

at the output. The current feedback circuits shown in Fig. 10.9(*c*) and (*d*) cause an *increase* in output impedance, since additional impedance is connected in series with the output to sense the output current. The series feedback circuits shown in Fig. 10.9(*a*) and (*c*) cause an *increase* in input impedance, since the feedback is connected in series with the input. Finally, the shunt feedback circuits shown in Fig. 10.9(*b*) and (*d*) cause a *decrease* in input impedance, since the feedback is connected in parallel with the input circuit.

Practical Exercise 10c

Negative feedback amplifier

Connect up the circuit shown in Fig. 10.10 without the feedback loop PQ, and resistor R_1 shorted out, on the matrix board. Switch on the supply, connect a sine wave input signal from the signal generator and monitor the input and output signal. Adjust the input signal to produce the maximum undistorted output. Determine the gain of the amplifier, and observe the effect on gain as the load resistance is varied.

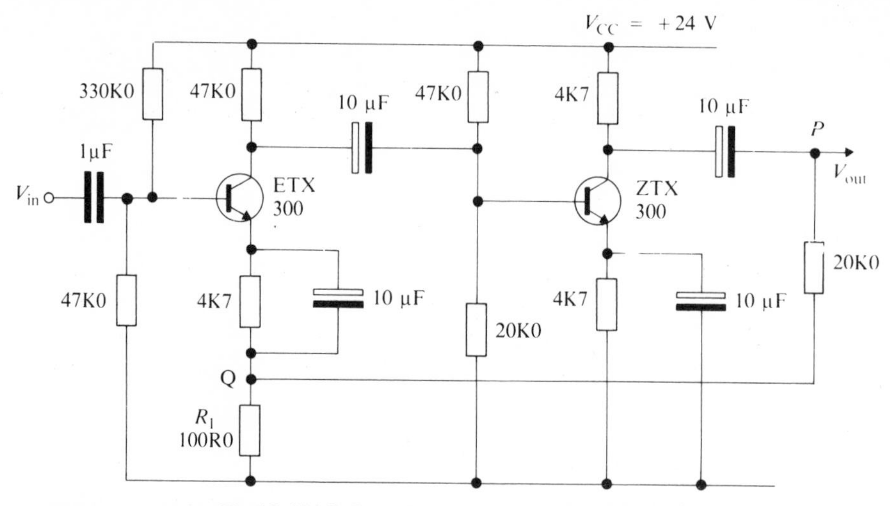

Fig.10.10 Series voltage negative feedback amplifier.

Apply a 'sweep' input signal, and sketch the gain/frequency response of the amplifier. Comment on this response.

Now, connect the feedback loop, together with R_1 to apply series voltage negative feedback, and repeat the above measurements.

Practical Exercise 10d

Positive feedback – RC phase shift oscillator

Connect up the circuit shown in Fig. 10.11, on the matrix board. With PQ

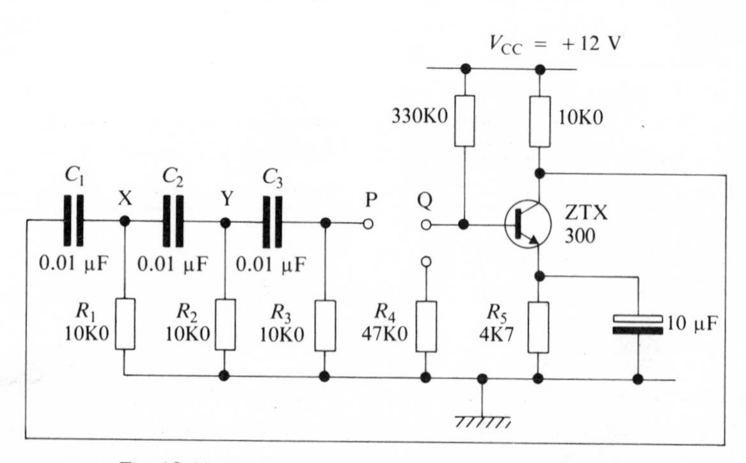

Fig. 10.11 Positive feedback – RC phase shift oscillator.

disconnected and R_4 connected to the transistor base, apply a 1 kHz sine wave to the input (Q). Monitor the output and determine the gain of the amplifier.

Note: The gain of the amplifier must be greater than 30. If not, replace the 10K0 collector resistor by a 25K0 variable resistor and adjust to produce the desired gain.

Disconnect the input signal.

Connect PQ and disconnect R_4. Determine the amplitude and frequency of the output signal.

Comment on the phase of the signal at X, Y, and P relative to output signal.

Disconnect PQ and connect R_4. Apply a sine wave signal to the input at Q. Adjust the signal to produce an output of the same amplitude and frequency as the oscillations. Monitor the output at point P and comment on the phase and amplitude relative to the input signal.

10.7 Operational amplifiers

Operational amplifiers were originally developed from differential amplifiers, which were used to compare two input signals. The circuit layout of transistorized operational amplifiers (op.amps) has made them ideal for integration – so that many types are available in IC packages. Consider an internally compensated op.amp, such as the SN 72741 (commonly referred to as the 741) which may be compared to the simple, single transistor amplifier, as shown in Fig. 10.12.

Both of these amplifiers require only five connections for inputs, outputs, and power supplies, but the op.amp. has marked advantages in most parameters, e.g., d.c. gain in excess of 200 000 compared with approximately 100 for the transistor amplifier, input impedance of the order of 2 MΩ compared with

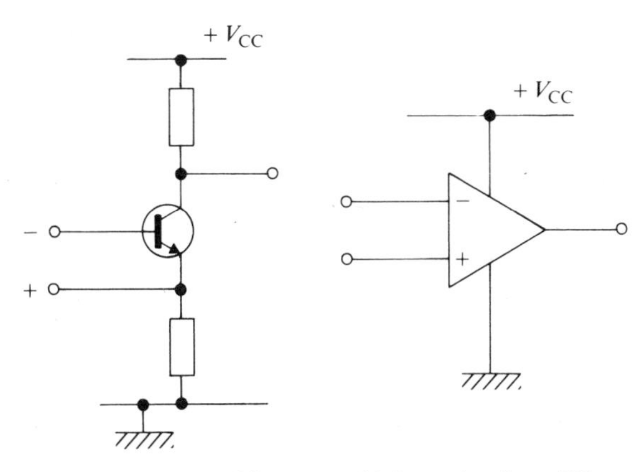

(*a*) Transistor amplifier (*b*) Operational amplifier

Fig. 10.12 Comparision of transistor amplifier and operational amplifier.

approximately 20 kΩ, and output impedance of 100 Ω compared with approximately 10 kΩ. Even the cost of an IC op.amp may be more advantageous.

The *ideal* op.amp. parameters are as follows: infinite gain, infinite bandwidth, infinite input impedance, zero input current, zero offset, and zero output impedance.

Most amplifiers have an input stage consisting of a pair of bipolar transistors in some form of long tailed pair. These will, of course, require a finite base current to keep them biased on and although the input transistors are very well matched, it is not possible to match them perfectly. Therefore, there will be a small input voltage offset and input current offset, V_{IO} and I_{IO}. Similarly, the differential input impedance between the input bases will be less than infinity and the output impedance of the amplifiers will be greater than zero.

For d.c. amplifiers, the most important parameters are usually the input offset voltage and current. To reduce the effect of the input bias current, a resistor R_3, as shown in Fig. 10.13, is included so that both amplifier terminals see exactly the same resistance to ground.

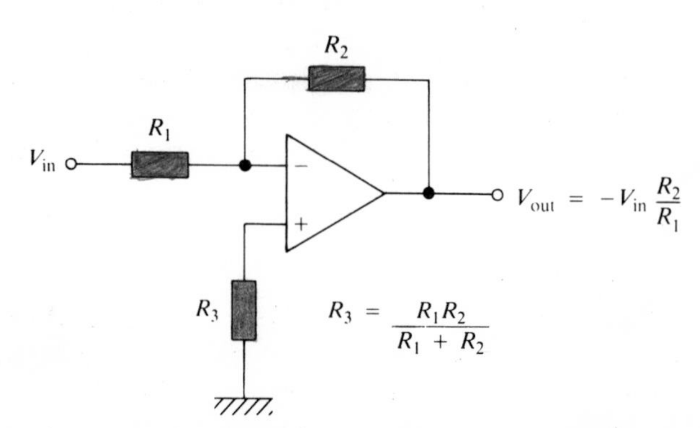

Fig. 10.13 Arrangement to reduce input bias current.

As the bias current to each terminal is sourced by the same impedance, there will be no offset produced at the inputs due to this current. There will, however, be a small output offset due to V_{IO} and I_{IO}.

10.8 Operational amplifier terminology

1. ***Input offset voltage*** (V_{IO}). The d.c. voltage which must be applied between the input terminals to force the quiescent d.c. output voltage to zero.
2. ***Input offset current*** (I_{IO}). The difference between the currents into the two input terminals with the output at zero volts.
3. ***Input bias current*** (I_{IB}). The average of the current into the two input terminals with the output at zero volts.

4. ***Input voltage range*** (V_I). The range of voltage which, if exceeded at either input terminal, will cause the amplifier to cease functioning properly.
5. ***Maximum peak-to-peak output voltage swing*** (V_{OPP}). The maximum peak-to-peak output voltage which can be obtained without waveform clipping when the quiescent d.c. output voltage is zero.
6. ***Large signal differential voltage amplification*** (A_{VD}). The ratio of the peak-to-peak output voltage swing to the change in differential input voltage required to drive the output.
7. ***Input resistance*** (r_i). The resistance between the input terminals with either input grounded.
8. ***Output resistance*** (r_o). The resistance between the output terminal and ground.
9. ***Input capacitance*** (C_I). The capacitance between the input terminals with either input grounded.
10. ***Common mode rejection ratio (CMRR).*** The ratio of differential voltage amplification to common-mode amplification – measured by determining the ratio of a change in input common-mode voltage to the resulting change in output offset voltage referred to the input.
11. ***Slew rate (SR).*** The average time rate of change of the closed-loop amplifier output voltage for a step-signal input. Slew rate is measured between specified output levels (0 V and 10 V) with feedback adjusted for unity gain.

Practical Exercise 10e

Op.amp. – offset adjust

Before an operational amplifier is used, its offset must be adjusted to zero, i.e., offset null. Offset is a measure of the output voltage when the input is zero. The connections for the 741 IC op.amp. are as shown in Fig. 10.14(*a*).

Connect up the circuit shown in Fig. 10.14(*b*). Switch on the supply and adjust the 10K0 variable resistor to give zero output voltage with zero input voltage.

Note: This setting should be made before all the op.amp. exercises.

Practical Exercise 10f

Op.amp. – fixed gain (inverting) amplifier

Adjust offset null as described in Practical Exercise 10e. Connect up the circuit as shown in Fig. 10.15(*a*).

1. Apply a d.c. voltage to the input terminal, and measure the output voltage.

Note: A positive input voltage will produce a *negative* output voltage. Also the output voltage *saturates* before it reaches $\pm V_{CC}$, therefore it will be necessary to restrict your input voltage to about 2.5 V in this case, since the gain is 100/20 = 5.

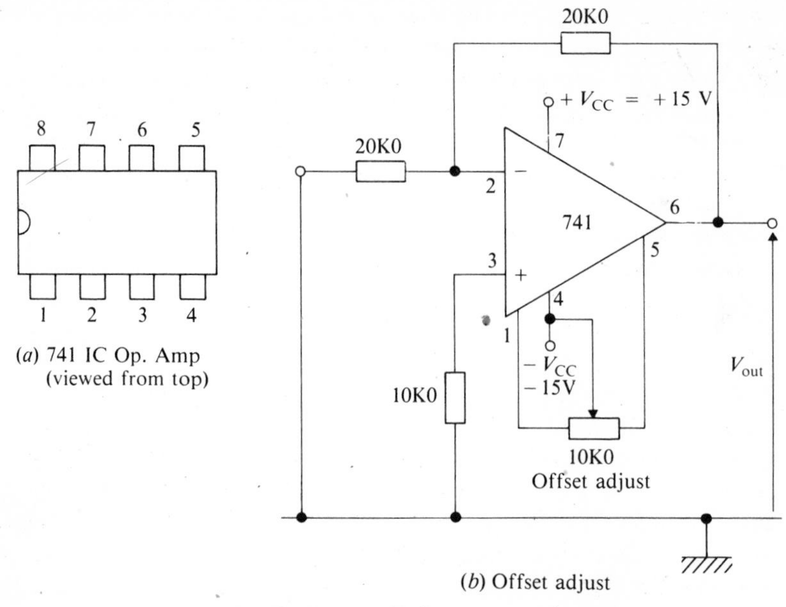

(b) Offset adjust

Fig. 10.14 The 741 Op. amp – setting up.

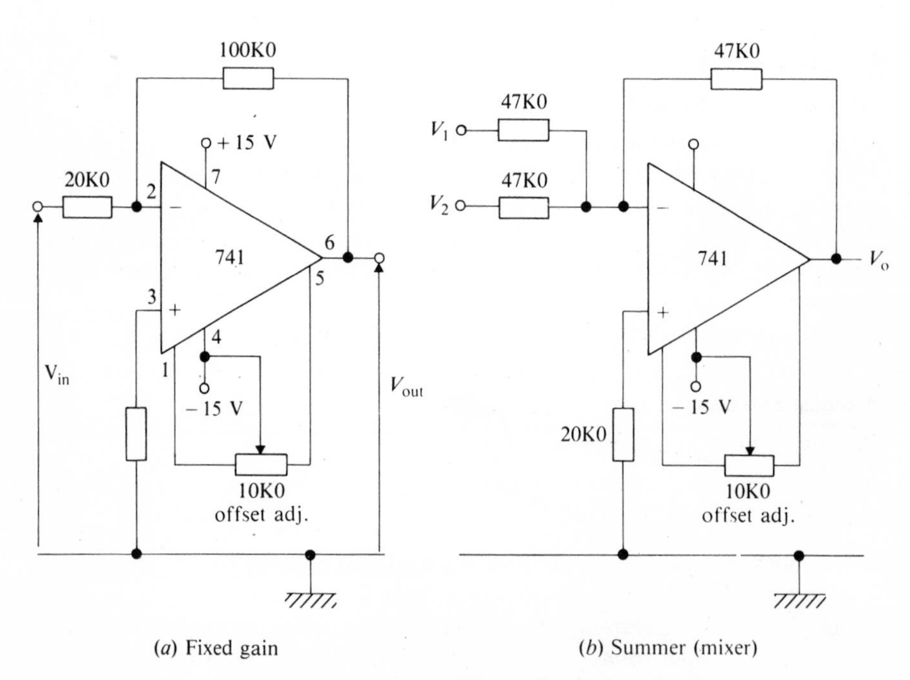

(a) Fixed gain

(b) Summer (mixer)

Fig. 10.15 Inverting amplifiers – fixed gain and summer.

Vary the input voltage through the range −3 V to +3 V to determine the saturation voltage in both the positive and negative directions.

2. Apply a 1 kHz, 1 V peak-to-peak sine wave to the input and measure the input and output using a dual-trace CRO. Determine the voltage gain. Compare with − 100K0/20K0. Change R_2 from 10K0 to 20K0 and repeat the above tests.

Practical Exercise 10g

Op.amp. – summer (mixer)

Adjust offset null as described in Practical Exercise 10e. Connect up the circuit shown in Fig. 10.15(*b*).

1. Apply two d.c. voltages to v_1 and v_2 and measure the output. Compare the result with $-(v_1 + v_2)$.
2. Apply two sine waves, at 1 kHz to v_1 and v_2 and measure the amplitude and frequency of the output. Change one of the input frequencies and observe the output. This is the arrangement for 'mixing' signals.

Change the resistor values to change the gain applied to each input signal and repeat the above measurements.

Practical Exercise 10h

Op.amp. – unity gain buffer amplifier

Connect up the circuit shown in Fig. 10.16(*a*).

Apply a d.c. voltage to the input and measure the output. This will be the same polarity and magnitude as the input.

Repeat this test with an a.c. input signal.

Practical Exercise 10i

Op.amp. – fixed gain (non-inverting) amplifier

Adjust offset null as described in Practical Exercise 10e. Connect up the circuit as shown in Fig. 10.16(*b*).

1. Apply a d.c. voltage to the input and measure the output voltage. Compare the output voltage with

$$\frac{R_1 + R_2}{R_1} v_{in}$$

2. Apply an a.c. signal sine wave at 1 kHz to the input and measure the output signal using the CRO. Compare the output voltage with

$$\frac{R_1 + R_2}{R_1} v_{in}$$

Repeat with different values of R_1 and R_2.

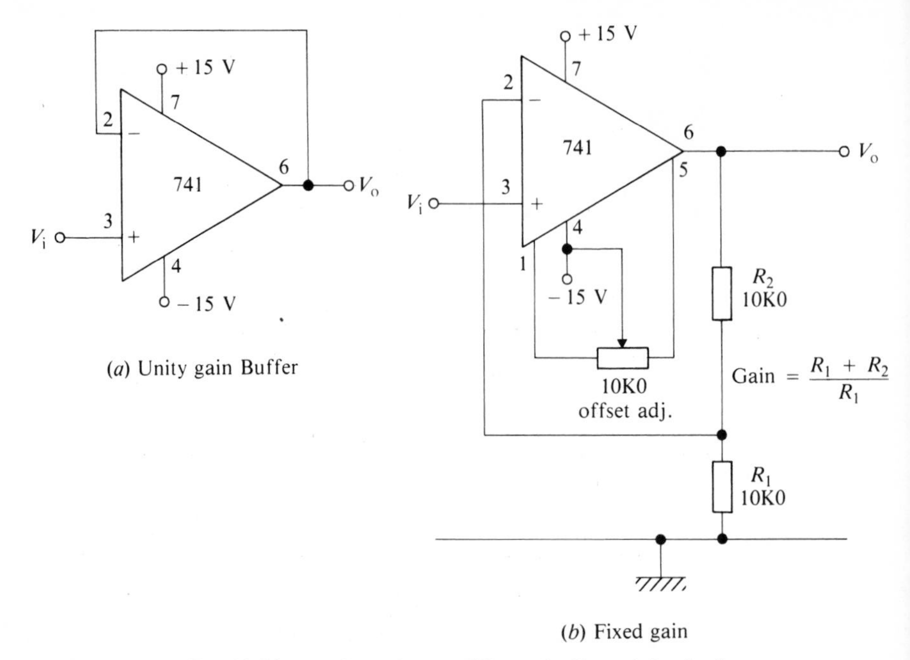

Fig. 10.16 Non-inverting amplifiers – buffer and fixed gain.

Practical Exercise 10j

Op.amp. – comparator (difference) amplifier

Adjust offset null as described in Practical Exercise 10e. Connect up the circuit shown in Fig. 10.17. Measure the output voltage for different combinations of the two d.c. input voltages. The output voltage is given by:

$$v_o = -R_2 \left(\frac{v_1}{R_1} - \frac{v_2}{R_4} \right) \text{ volts.}$$

Practical Exercise 10k

Op.amp. – integrator

Adjust offset null as described in Practical Exercise 10e. Connect up the circuit as shown in Fig. 10.18(*a*).

1. ***d.c.*** – With R = 100K0 and C = 100 μF, apply a d.c. signal to the input, and observe the output voltage change until saturation is reached. Repeat with different values of R and C.
2. ***a.c.*** – With R = 10K0 and C = 0.01 μF, apply an a.c. square wave at 1 kHz to the input. Observe the output voltage waveshape and amplitude using the CRO. Repeat with different values of R and C.

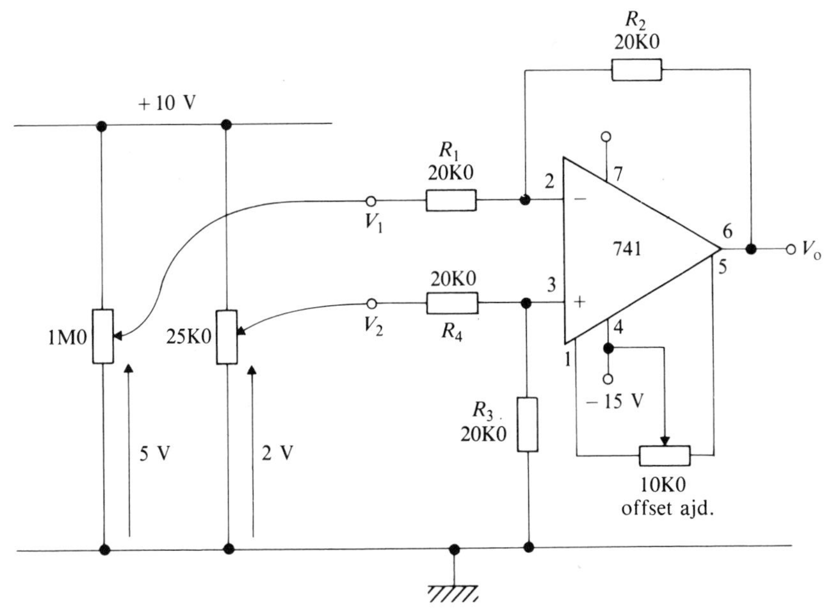

Fig. 10.17 Op. amp comparator.

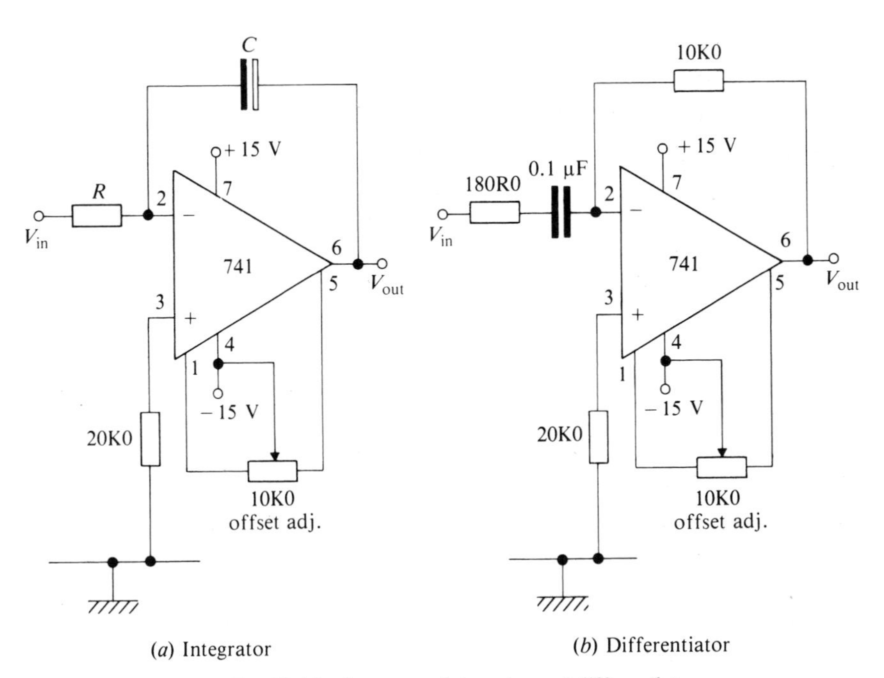

(*a*) Integrator (*b*) Differentiator

Fig. 10.18 Op. amp – integrator and differentiator.

Practical Exercise 10l

Op.amp. – differentiator

Adjust offset null as described in Practical Exercise 10e.Connect up the circuit shown in Fig. 10.18(*b*). Apply a square wave at 100 Hz to the input and observe the output waveform using the CRO.

Practical Exercise 10m

Op.amp. – triangle and square wave oscillator

Adjust offset null for both op.amps as described in Practical Exercise 10e. Connect up the circuit shown in Fig. 10.19. This is a Schmitt trigger followed by an integrator. Observe the output waveforms: square wave at A, triangle at B, as the frequency adjust is varied.

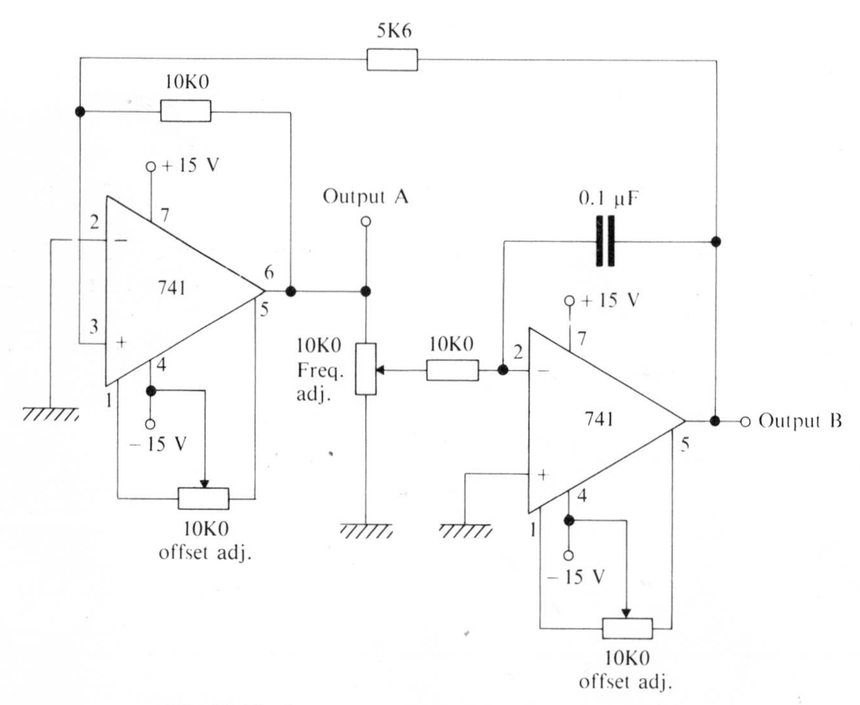

Fig. 10.19 Op. amp – triangle and square wave oscillator.

10.9 The power amplifier

We have considered the conditioning of a low-level signal so far, but at some stage in most industrial systems it is necessary for some work to be done in response to the signal. This inevitably involves power, so that it is necessary to convert the voltage signal into a power output – this is achieved in the power amplifier.

Consider the arrangement shown in Fig. 10.20.

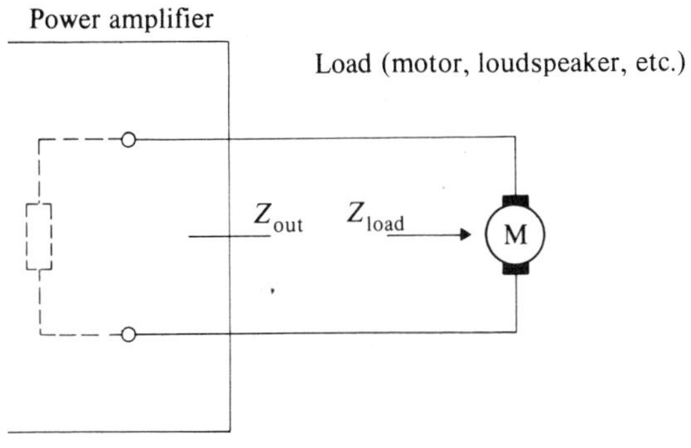

Fig. 10.20 Maximum power transfer.

Maximum power is transferred from a source to a load when

$$Z_{out} = Z_{load}.$$

In power amplifiers, it is, therefore, necessary to match these impedances to achieve maximum efficiency of operation.

The power amplifier is basically the same as a voltage amplifier except that it handles higher currents and generally supplies a low load impedance, such as a loudspeaker or a motor.

The desired degree of impedance matching can be achieved using a transformer, as shown in Fig. 10.21. The transformation ratio n is chosen by:

$$n = \sqrt{Z_{out}/Z_{load}}.$$

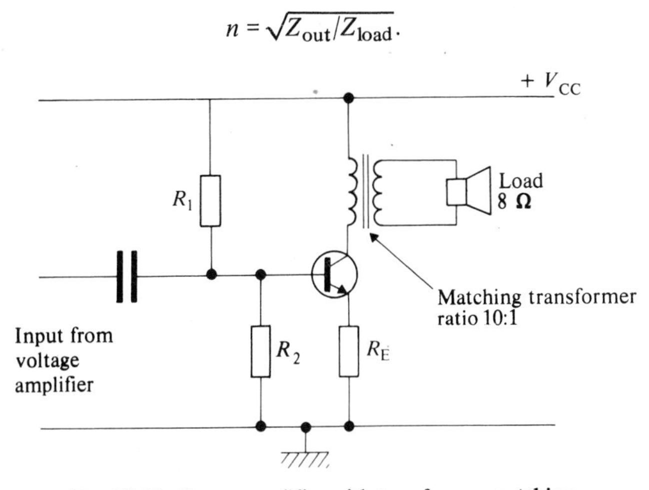

Fig. 10.21 Power amplifier with transformer matching.

The circuit shown in Fig. 10.21 is referred to as *Single-ended* (since only *one* transistor is used). *Class A power amplifier*. This circuit may include a n.t.c. thermistor in parallel with R_2 to ensure that thermal stability is maintained.

When more power is required than that which can be provided from single-ended power amplifiers, the *Class B push-pull power amplifier* may be used as shown in Fig. 10.22. The circuitry is arranged so that each transistor amplifies half of the signal, so that twice the power can be handled, and no current flows in the absence of a signal.

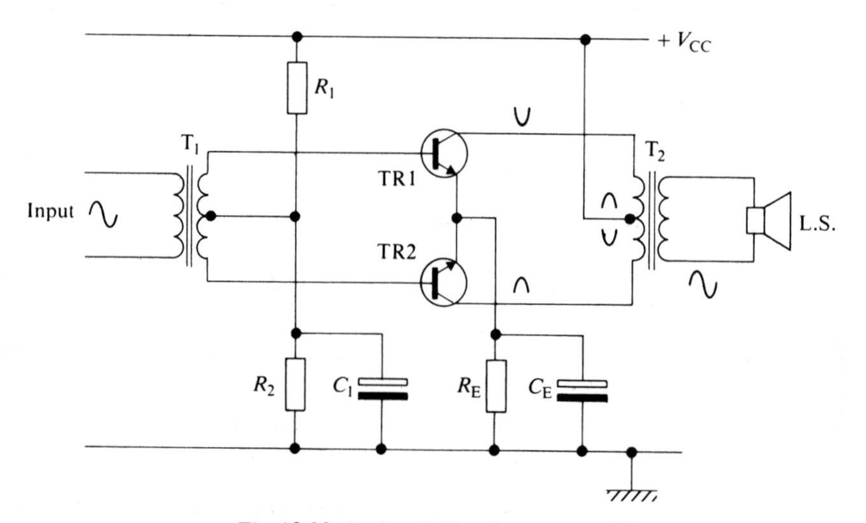

Fig. 10.22 Push-pull Class B power amplifier.

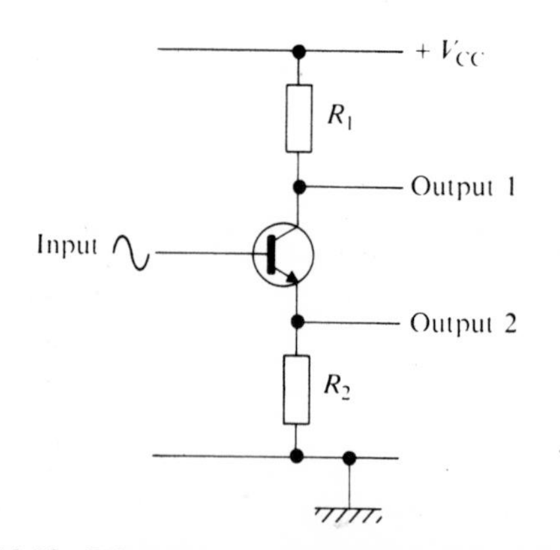

Fig. 10.23 Driver stage for push-pull power amplifiers.

10.10 Driver stage for push-pull power amplifiers

The disadvantage of the driver transformer T_1 (Fig. 10.22) may be overcome by using a transistor driver (or *phase splitter*) as shown in Fig. 10.23. Output 1 uses R_1 as its load and is in anti-phase with the input due to the inherent phase reversal between base and collector. Output 2 uses R_2 as its load, and is in phase with the input signal, i.e., emitter follower.

The transistor driver is used in the complementary Class B push-pull power amplifier shown in Fig. 10.24. This may also be referred to as a transformerless push-pull amplifier. Transistors TR_2 and TR_3 are *complementary* transistors, i.e., one is a n p n and the other is p n p – 'matched' to have the same gain and other properties as close as possible.

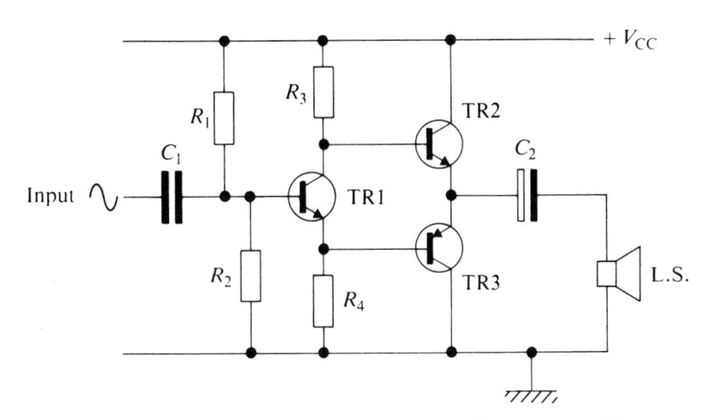

Fig. 10.24 Complementary Class B push-pull power amplifier.

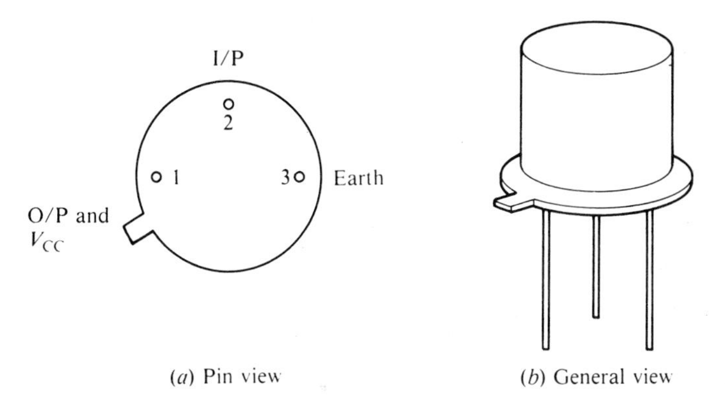

Fig. 10.25 The Ferranti ZN414 a.m. radio chip.

Practical Exercise 10n

Amplitude-modulated radio receiver

The Ferranti ZN 414 is a 'single-chip' 10-transistor tuned radio receiver, packaged in a 3-pin transistor-outline (TO18) case, as shown in Fig. 10.25. Pin 2 is the radio-frequency input, pin 1 is the audio signal output (also used for the supply) and pin 3 is connected to ground. The ZN414 fulfils the requirements of a tuned r.f. stage and detector. The audio power amplifier stage can be provided using the LM380 2 W amplifier feeding an 8 Ω loudspeaker.

Connect up the circuit shown in Fig. 10.26, where VC_1 is a variable air-dielectric capacitor of 500 pF and the coil L may be constructed by winding 55 turns of insulated copper wire around a rod of ferrite. The size of the wire is not critical, but about 36 s.w.g. should be suitable.

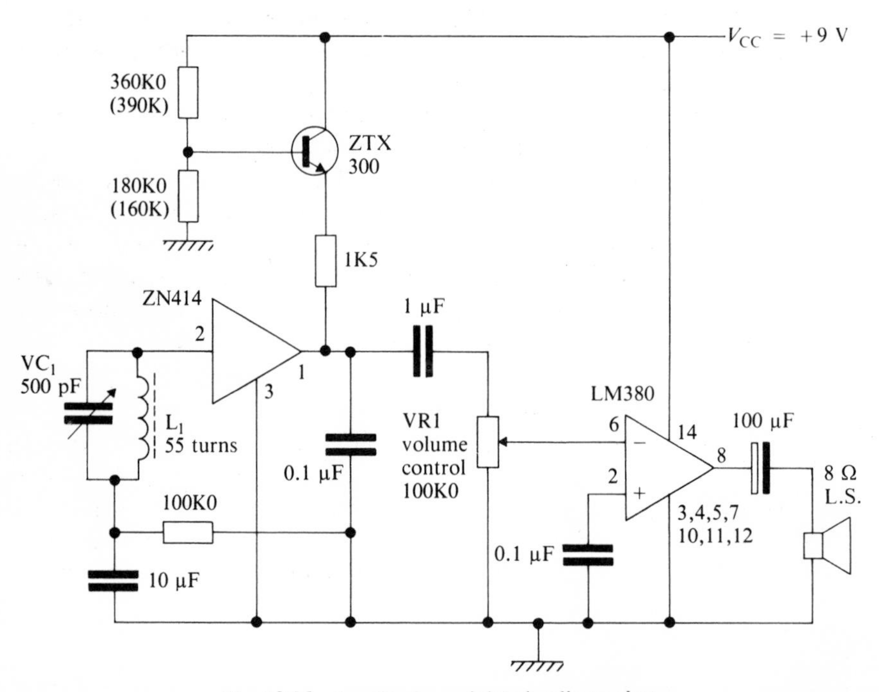

Fig. 10.26 Amplitude modulated radio receiver.

Switch on the supply, turn up the volume control and tune the radio to a suitable programme by adjusting the variable capacitor VC_1.

If the results are not entirely satisfactory try changing the ZTX 300 biasing components to those indicated in brackets in Fig. 10.26 and repeat the test.

11. Optoelectronics

11.1 Introduction

During the early development of semiconductors, it was discovered that diodes and transistors were sensitive to light and that it was necessary to use opaque materials in the encapsulation to enable the devices to operate correctly. It had also been observed that some devices emitted visible light when operated under certain conditions. The introduction of silicon light sensors (about 1950) and light-emitting diodes (about 1960) with their high production costs restricted the use of these devices to specialized applications. More recently, however, the developments made in semiconductor technology and production techniques have led to the availability of a wide range of low cost optoelectronic devices which are finding increasing use in many industrial applications.

Semiconductor optoelectronic devices are basically diodes and transistors which are specially designed to make efficient use of the fact that all semiconductor devices interact with light to some extent. Optoelectronic devices may be broadly divided into two groups: *light sensors* (or *detectors*) are diodes or transistors that convert light energy into electrical energy; *light emitters* (or *sources*) are diodes that convert electrical energy into light energy.

11.2 Light emitters

Light sources such as tungsten filament lamps, fluorescent lamps, and neon lamps are normally designed to emit visible light. Semiconductor light emitters are p n junction devices which emit light when they are forward biased, i.e., positive to p type (anode). These devices are called LEDs (light-emitting diodes) and the circuit symbol is shown in Fig. 11.1. Since LEDs are p n junction diodes,

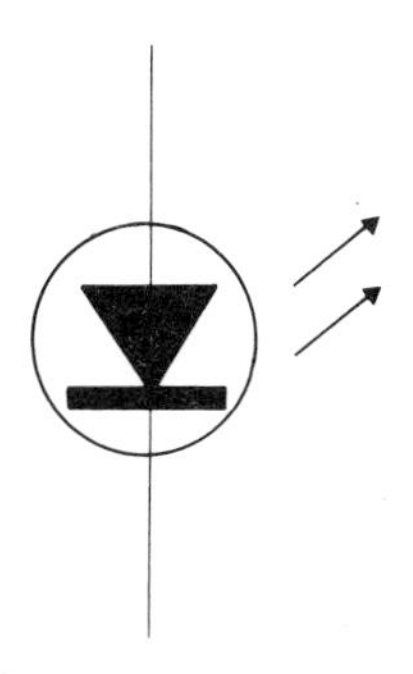

Fig. 11.1 Circuit symbol for the LED.

they have the same electrical characteristics as a normal diode, i.e., they conduct current when forward biased and block the flow of current when reverse biased, but they also produce light energy (in the form of *photons*) efficiently when forward biased.

Semiconductor light emitters are made in a wide range of wavelengths and, hence, a range of colours is theoretically possible. In practice, however, manufacture is generally limited to the spectral response of silicon, gallium arsenide, gallium phosphide, and gallium arsenide phosphide; the most widely available colours are red, green, yellow, and orange in the visible range. In many electronic applications the LEDs are infrared-emitting diodes (IRED), the emissions of which are invisible to the human eye. However, the majority of semiconductor light sensors are most sensitive to infrared light, which has led to a wide range of interesting applications.

Visible-light-emitting diodes (VLEDs) are widely used as indicators and as digital displays in instruments, electronic calculators, digital clocks, and watches. VLEDs are usually intended to be operated with a series resistor, as shown in Fig. 11.2(*a*), to limit the current. Typical values of I_F are 5 and 25 mA (red) and a forward voltage of about 2 V. Check with manufacturer's data to be sure. Terminal identification varies with the manufacturer, as shown in Fig. 11.2(*b*). Again, check with manufacturer's data before use.

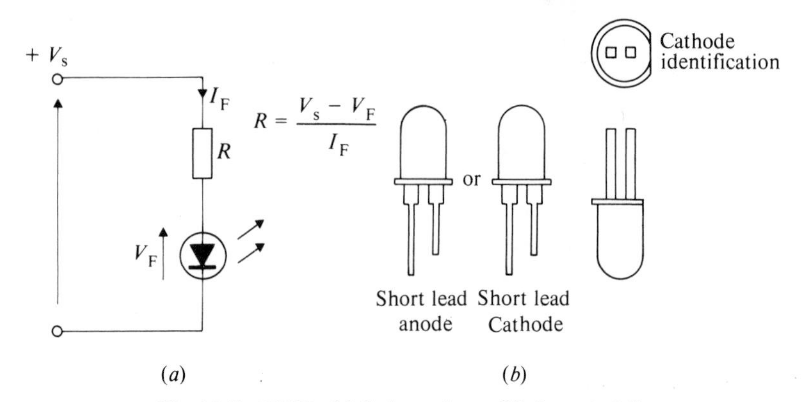

Fig. 11.2 VLEDs (*a*) Series resistor; (*b*) characteristics.

Visible displays are available in a wide range of different forms; a typical seven-segment numerical display is shown in Fig. 11.3. In addition to the display shown, alpha-numeric displays using both the seven-segment form and a dot matrix arrangement of VLEDs, e.g., a 5 x 7 dot matrix, are becoming more widely used.

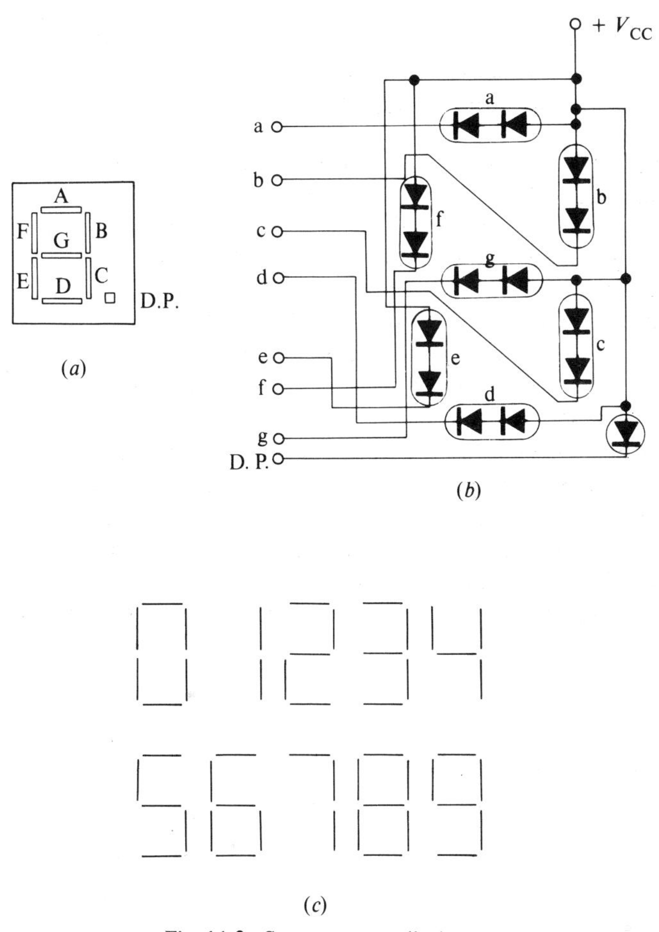

Fig. 11.3 Seven-segment display.

11.3 Light sensors

Semiconductor light sensors experience a change in their electrical characteristics when light energy releases charge carriers in the material, thus causing a change in its conductivity. Apart from the photoresistor, which we considered in Chapter 2, the main light detecting devices are:

1. photo-diode,
2. photo-transistor, including photo-FET,
3. photo-Darlington,
4. photo-thyristor.

11.4 The photo-diode

The photo-diode is a p n junction which is designed to operate when it is reverse biased, as shown in Fig. 11.4.

When light energy of the correct wavelength falls on the photo-diode junction, current is caused to flow in the external circuit. This device then acts as a current generator in which the current is directly proportional to the light intensity. Silicon is the most widely used material for photo-diodes, and gives response times of the order of 1 ns.

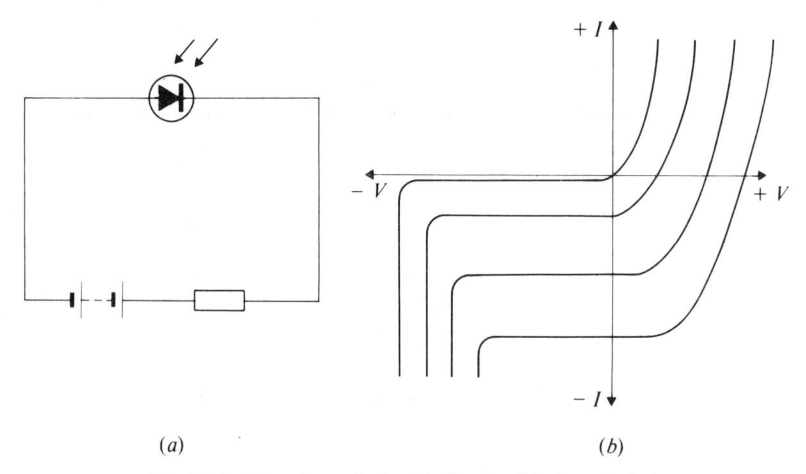

Fig. 11.4 The photo-diode. (*a*) Circuit; (*b*) characteristics.

11.5 The photo-transistor

The photo-transistor operates exactly like a normal transistor, except that instead of supplying an external base current to drive the transistor, the *photo-diode* existing between the collector and base is used as a current source. The circuit symbols for the photo-transistor are as shown in Fig. 11.5(*a*), in which the base connection may or may not be brought out to a terminal, and the equivalent circuit of the photo-transistor is shown in Fig. 11.5(*b*). The transistor effectively amplifies the photo-diode current by its current gain h_{FE}.

The dark current of the photo-diode is also amplified by h_{FE}, which means that the photo-transistor leakage current is higher than in conventional silicon transistors. When the photo-transistor is used to detect very low intensity light levels, the effects of the dark current can be reduced by maintaining a slight forward bias on the collector–base junction, as shown in Fig. 11.6.

The junction gate field effect transistor (JUGFET) has also been utilized to make a photo-FET light detector. The drain–gate junction in an *n* channel

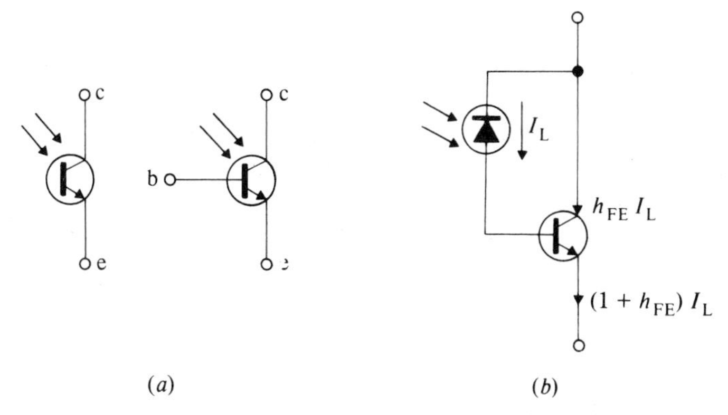

Fig. 11.5 The photo-transistor. (*a*) Circuit symbols; (*b*) equivalent circuit.

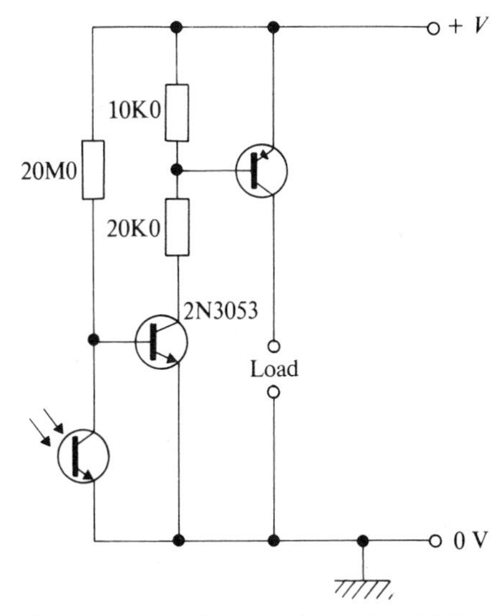

Fig. 11.6 Photo-transistor circuit to detect low light intensities.

FET is usually reverse biased – so this p n junction is used to produce a photocurrent which, if passed through a resistor R_G, as shown in Fig. 11.7, produces a gate voltage which is then amplified in the usual way.

11.6 The photo-Darlington

The photo-Darlington is basically the same as the photo-transistor, except that a very much higher gain is possible (typically $h_{FE_1} \times h_{FE_2} = 10\,000$) due to

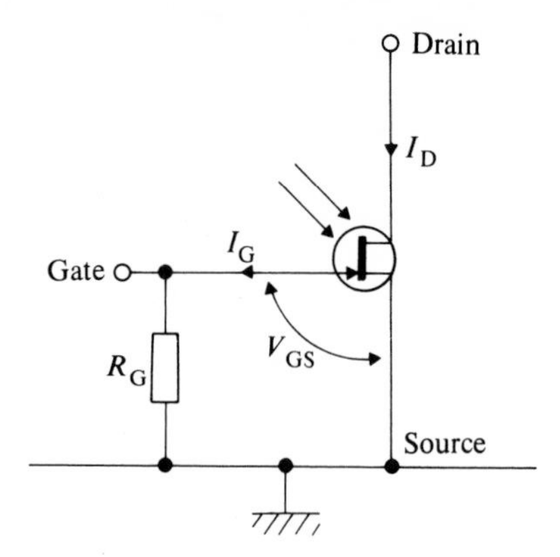

Fig. 11.7 The photo-FET.

the cascaded transistor connection. The circuit symbol is shown in Fig. 11.8(*a*) and the equivalent circuit is shown in Fig. 11.8(*b*).

11.7 The photo-thyristor

The photo-thyristor is similar to a normal p n p n silicon thyristor except that the area of one collector junction is increased to enable the device to be triggered by light energy. The circuit symbol of the photo-thyristor is shown in Fig. 11.9.

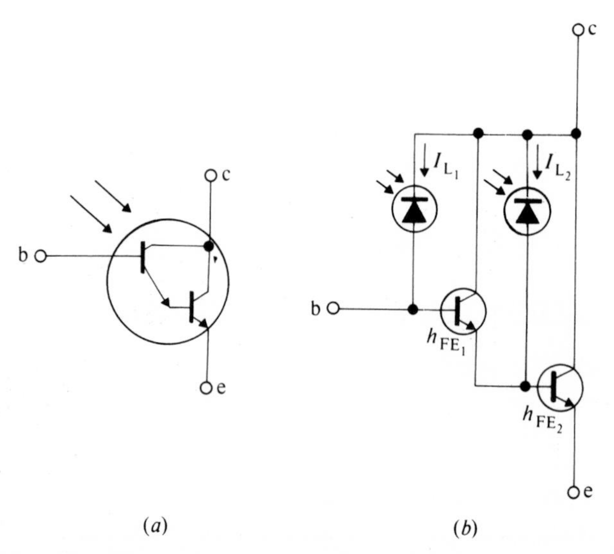

Fig. 11.8 The photo-Darlington. (*a*) Circuit symbol; (*b*) equivalent circuit.

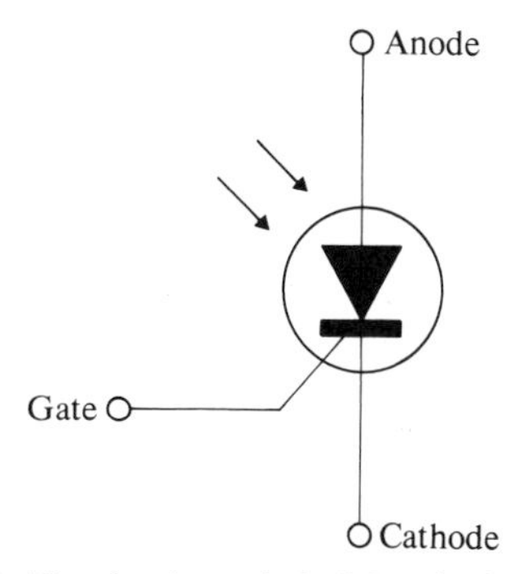

Fig. 11.9 The circuit symbol of the photo-thyristor.

11.8 Emitter/detector combinations

A wide range of devices utilizing the combined effects of both light emitters and light sensors is now readily available. These devices are called *opto-interrupters, opto-reflectors*, and *opto-couplers* (or *opto-isolators*), are commercially available in plastic packages, and may incorporate a lens and/or a filter to increase sensitivity. The arrangements commonly used for the interrupter and reflector devices are illustrated in Fig. 11.10.

The opto-coupler or opto-isolator is a purely electronic package, in which the path of light – usually infrared – is completely enclosed in the package. This provides a one-way transfer of electrical energy, from the IRED to the

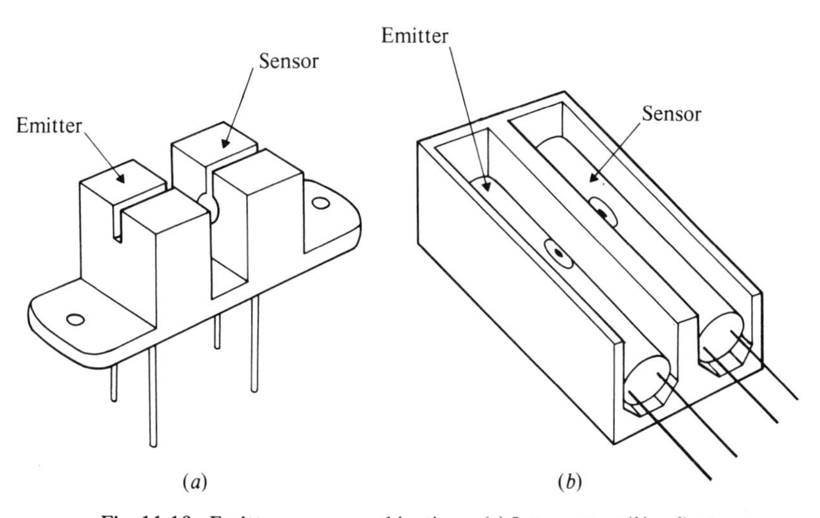

Fig. 11.10 Emitter-sensor combinations. (*a*) Interrupter; (*b*) reflector.

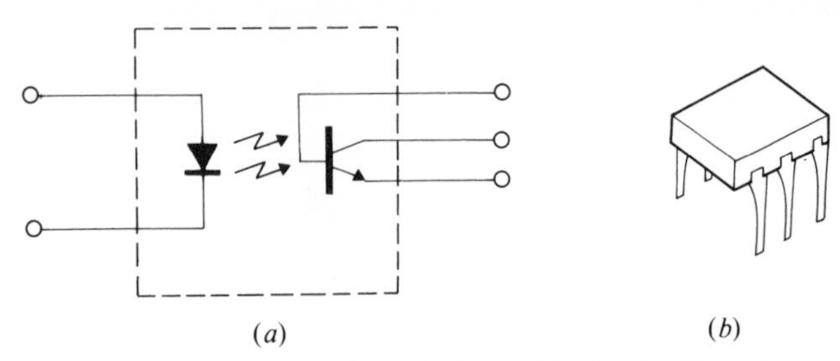

Fig. 11.11 Optocoupler (opto-isolator). (*a*) Circuit; (*b*) package.

photo-detector, whilst maintaining electrical isolation between the two circuits, as shown in Fig. 11.11.

11.9 Applications of optoelectronics

Many applications have been realized using the optoelectronic devices which we have discussed. Several circuits which utilize these devices are considered below:

1. ***Optical relay.*** As shown in Fig. 11.12, in which the photo-transistor and the n p n transistor are connected in a Darlington-pair connection to provide the relay coil current.

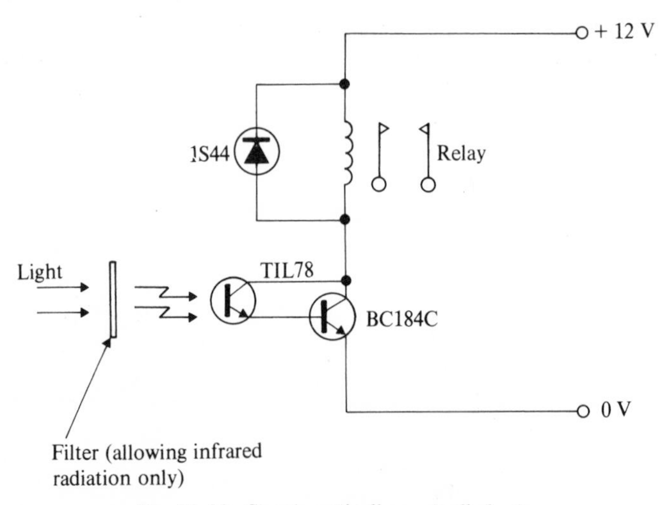

Fig. 11.12 Simple optically-controlled relay.

2. ***Automatic night-light.*** As shown in Fig. 11.13. This circuit is useful in a children's nursery, in which the lamp is automatically switched on when darkness occurs.

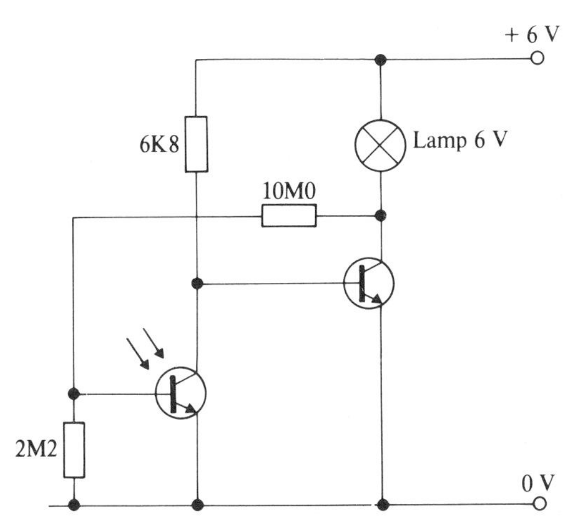

Fig. 11.13 Simple automatic battery night-light.

3. ***Automatic dipping of headlamps.*** As shown in Fig. 11.14. The phototransistor senses the lights of approaching traffic and, due to the 'courtesy' distance involved, the sensitivity of the detector is improved by the use of a lens. In addition, precautions must be taken to avoid the circuit being operated by flashing lights, and causing the lamps to switch from 'full beam' to 'dip' and back again at a rapid rate – these effects are restricted by building 'hysteresis' into the circuit.

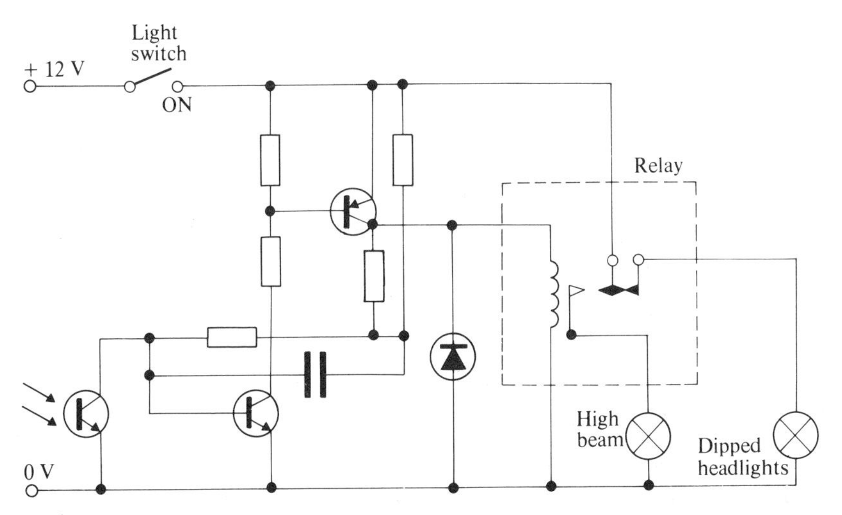

Fig. 11.14 Automatic dipping of car headlights.

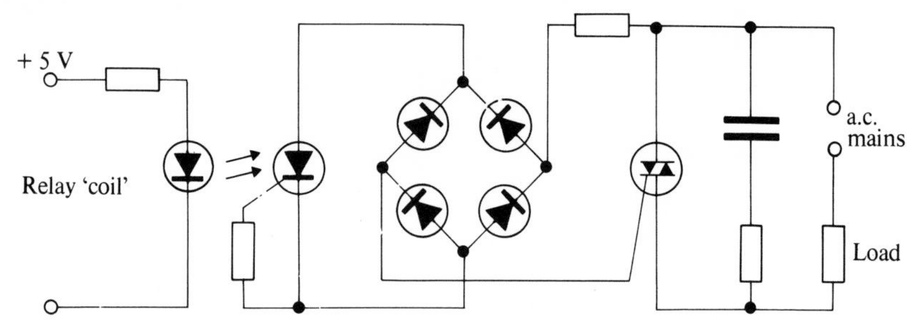

Fig. 11.15 Solid-state relay.

4. ***Solid-state relay.*** As shown in Fig. 11.15.
5. ***Pulse detector.*** As shown in Fig. 11.16.
6. ***Optical communications.*** As shown in Fig. 11.17. Both the transmitter and receiver circuits use 741 op. amps, and the system can be driven from a dynamic microphone at distances up to 50 m (between transmitter and receiver).
7. ***Production line flow.*** As shown in Fig. 11.18. Several optoelectronic methods can easily be used to control different aspects on the production

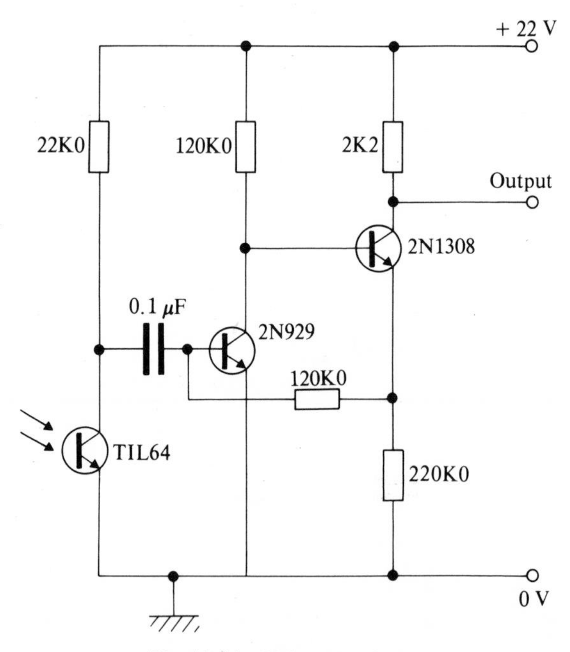

Fig. 11.16 Light pulse detector.

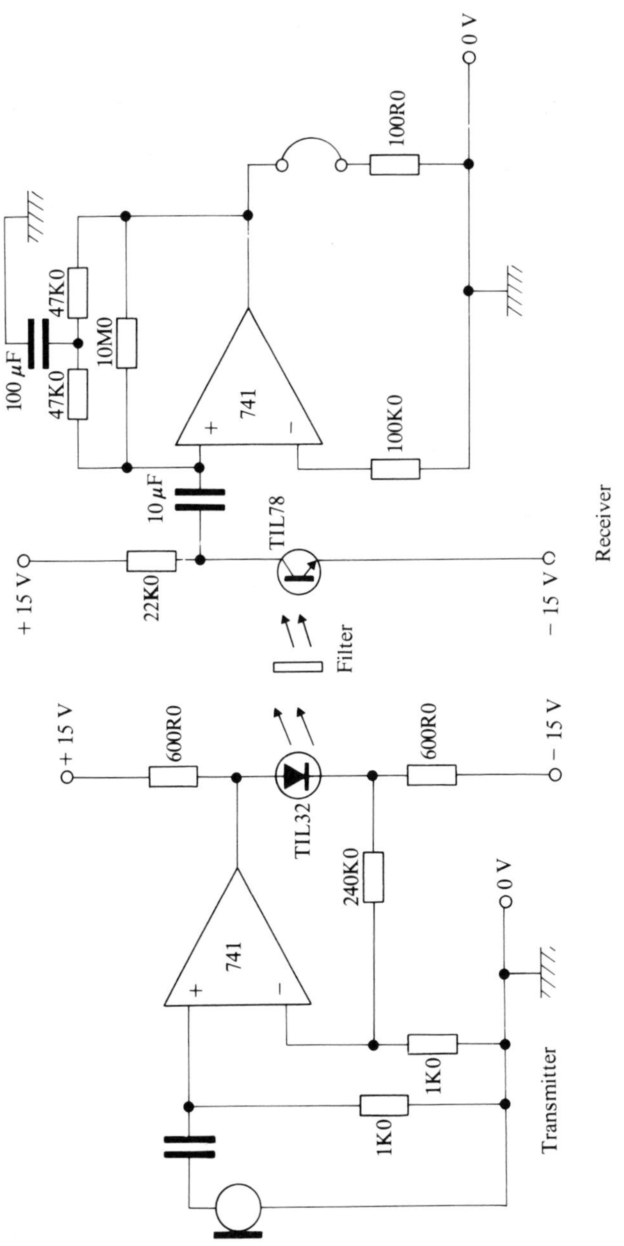

Fig. 11.17 Optical communications system.

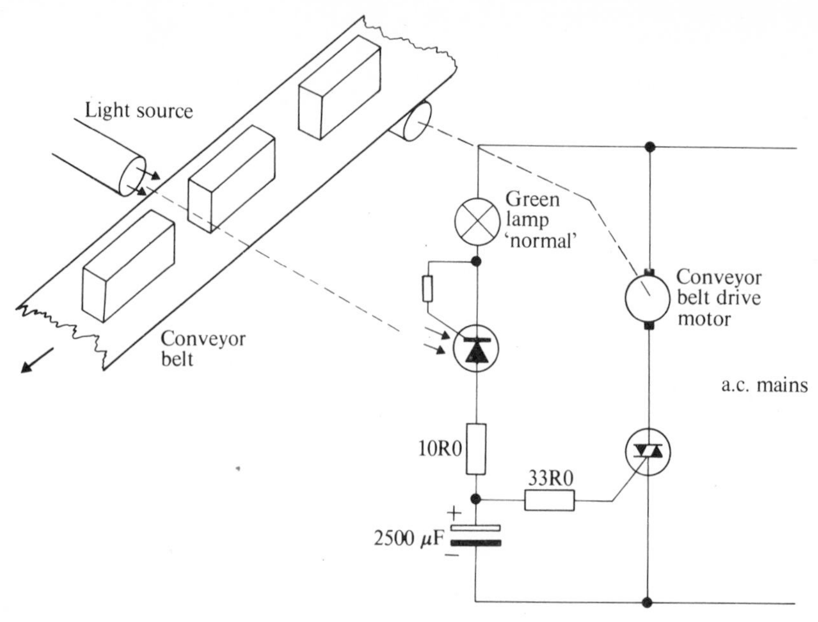

Fig. 11.18 Production line flow.

line. This system stops the conveyor belt drive motor if a blockage occurs in the flow of products, e.g., if the light source is blocked for longer than, say, 0.5 s, then the motor is switched off until the blockage has been cleared, when the triac is again triggered, allowing the motor to restart automatically.

Practical Exercise 11a

LED visible emitter – indicator lamp

Connect up the circuit arrangement shown in Fig. 11.19.

Measure the current flowing and the voltage across the LED and observe the light intensity emitted by the LED as the 2K0 variable resistance is adjusted from its maximum to its minimum value. *Do not exceed the rated values of the LED.*

Record the values of current and voltage and determine the value of series resistance to produce a useful indicator lamp.

Practical Exercise 11b

Optical pulse generator for electronic tachometer

Connect up the arrangement shown in Fig. 11.20. Switch on the supply and adjust the 'set speed potentiometer' until the motor runs at a reasonable speed. *Do not exceed 3000 rev min*$^{-1}$.

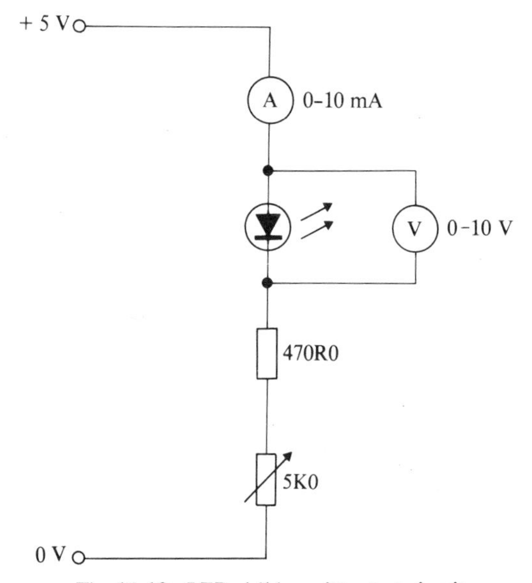

Fig. 11.19 LED visible emitter test circuit.

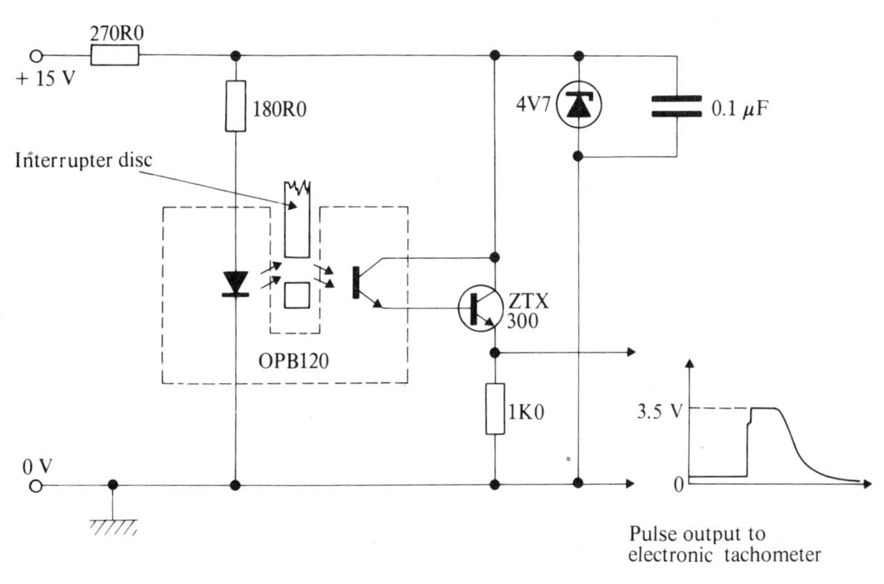

Fig. 11.20 Optical pulse generator for an electronic tachometer.

Connect the oscilloscope leads across the output of the pulse generating circuit and examine the pulse waveform in detail. Hence determine the motor speed. Set the motor speed to 1800 rev min^{-1}.

Systems of this type are particularly suited to motors in which the loading effect of a tachogenerator would be excessive.

In this particular system, a disc, having 10 holes around its edge, is mounted on the motor shaft such that it passes through an OPB 120 infrared emitter/detector arrangement. Rotation of the motor causes the disc to allow radiation through to the photo-transistor ten times each revolution. This has a pulsing effect in the photo-transistor and the circuit associated with it develops the voltage pulses which are fed into additional circuitry which is the electronic tachometer.

11.10 Displays

Three forms of display device have evolved as being the most popular, and therefore the most commonly encountered:

1. ***7-segment LED.*** This has been described in Sec. 11.2 and shown in Fig. 11.3. The 7-segment LED display will be used in Chapter 12 to display the 'count' output from digital counters, where 'special' circuits are used to energize the correct combination of segments.
2. ***5 x 7 dot matrix LED.*** There is an increasing requirement to display alphabet characters as well as numeric symbols. A matrix of LEDs is arranged as five columns of seven rows, as shown in Fig. 11.21 (*a*). The connections are arranged so that an LED is connected between each row wire and each column wire as shown in Fig. 11.21(*b*).

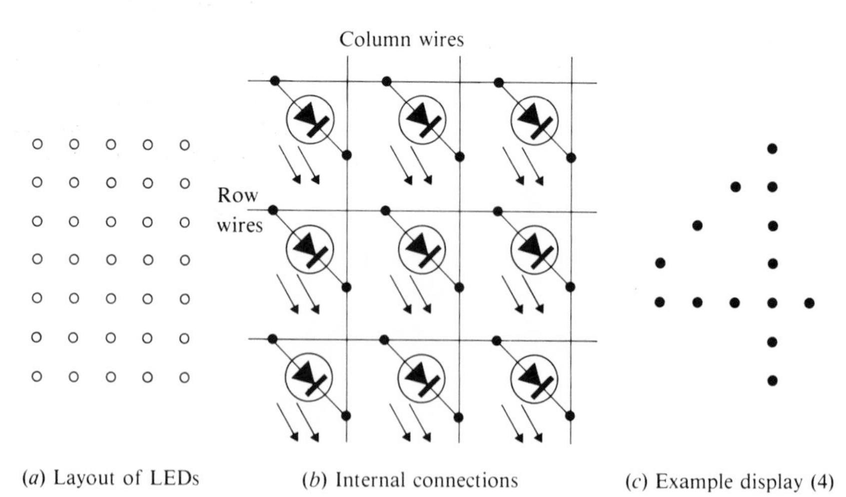

Fig. 11.21 5 x 7 dot matrix LED display.

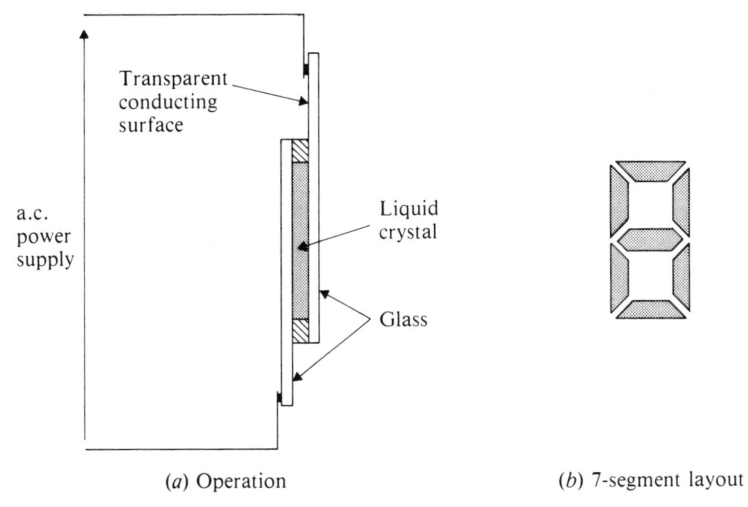

(*a*) Operation (*b*) 7-segment layout

Fig. 11.22 Liquid crystal display (LCD).

3. ***Liquid crystal, LCD.*** The principle of operation of the LCD is shown in Fig. 11.22(*a*), in which the liquid crystal is sealed between two glass surfaces, the inner surface of each being coated with a transparent conducting material. Two types exist: the field effect (usually producing a black display) and the dynamic scattering type (which generally produces a white display). The liquid crystal itself does not produce any illumination, but depends upon ambient light to enable the display to be observed. Liquid crystal displays suffer from reduced life if energized from d.c. supplies. Therefore, the crystals are energized by an a.c. supply, and require only a very small current (microamperes). Application of a potential (a.c.) in the range 1.5 V to 30 V cause the optical properties of the crystal to change, thus allowing the display to be observed through the segments as shown in Fig. 11.22(*b*).

12. Logic and Digital Electronics

12.1 Introduction

The subject of electronic logic is one which embraces the whole field of electronics from computers to automobiles and from toys to telephone exchanges. 'Logic' devices serve man in every walk of life, each device operating in a predictable manner. In fact, so predictable is their operation that we can use a form of *logical algebra* to determine the way in which a system works.

Control logic has developed to keep pace with advances in the electronics field. Without it, the construction of computers and electronic control equipment, leading to space flight and moon walks, would have been impossible.

12.2 Boolean algebra

George Boole studied 'logical thought' around the middle of the 19th century and established a new system called 'the algebra of classes'. Unfortunately, at that time there was very little interest in the application of this new system. It was not until just before the Second World War that Shannon discovered that Boole's algebra of classes was a very powerful method which was ideal to analyse and represent complicated circuitry employing 'two state' ideas.

The basic rules of Boolean algebra are:

1. A quantity can have only *one* of two possible values, it can be a '1' or a '0'. No other value exists.
2. The usual meaning of certain mathematical signs take on an entirely different meaning:

    $A.B$ means A *and* B (not A times B).
    $A + B$ means A *or* B (not A and B),
    $\bar{A}$ means *not* A.

3. The sign of equality (=) has a new significance which may best be shown as: = means 'an output exists' or 'the switch is closed'.

12.3 Logical elements

Devices used in logic networks control the flow of *information* through the system, and, for this reason, are referred to as *logic gates*, since the 'gates' are opened and closed by the combination or sequence of events occurring at their inputs. The basic range of gates are known by the names AND, OR, NOT, NOR, and NAND.

12.4 The AND gate

The AND gate may be represented by a number of series-connected switches as shown in Fig. 12.1(*a*). The logic symbol for this two input arrangement is shown in Fig. 12.1(*b*). The lamp will only be lit (logical '1') if both A and B are closed, i.e., both at logical '1'. The *truth table* shows the output state for all possible combinations of inputs, as shown in Fig. 12.1(*c*). Therefore, the truth table represents the Boolean equation

$$F = A.B.$$

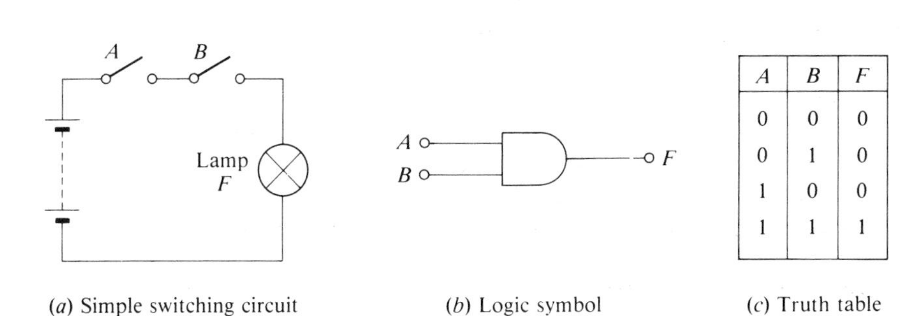

A	B	F
0	0	0
0	1	0
1	0	0
1	1	1

(*a*) Simple switching circuit (*b*) Logic symbol (*c*) Truth table

Fig. 12.1 The AND gate.

12.5 The OR gate

The OR gate may be represented by a number of parallel connected switches as shown in Fig. 12.2(*a*). In this case the lamp will be lit (logical '1') if A or B or both are closed, i.e., logical '1'. The logic symbol for this two input arrangement is shown in Fig. 12.2(*b*). The truth table is shown in Fig. 12.2(*c*), and the Boolean equation is

$$F = A + B$$

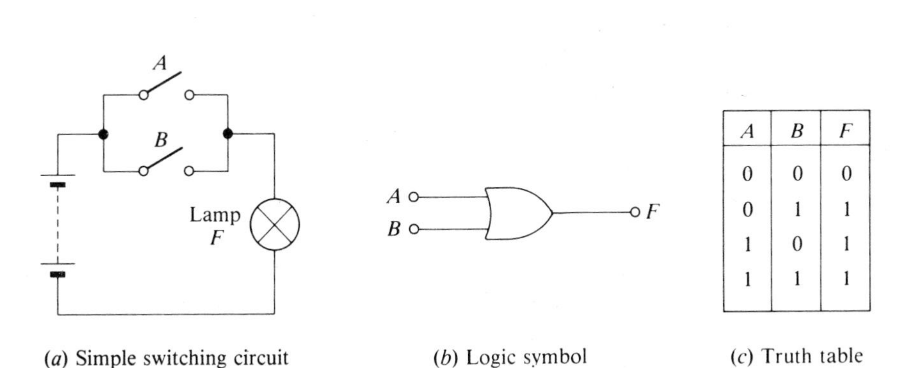

A	B	F
0	0	0
0	1	1
1	0	1
1	1	1

(*a*) Simple switching circuit (*b*) Logic symbol (*c*) Truth table

Fig. 12.2 The OR gate.

12.6 The NOT gate

This is frequently called an *inverter, negater*, or simply *sign changer.* The symbol is shown in Fig. 12.3. Note the small circle which indicates the change of sign.

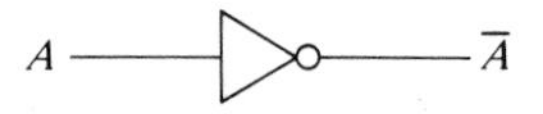

Fig. 12.3 The NOT gate.

12.7 The NOR gate

This is NOT-OR, and may be realized using an OR gate followed by a NOT gate. The symbol and truth table for the NOR gate are shown in Fig. 12.4. The Boolean equation is

$$F = \overline{A + B}$$

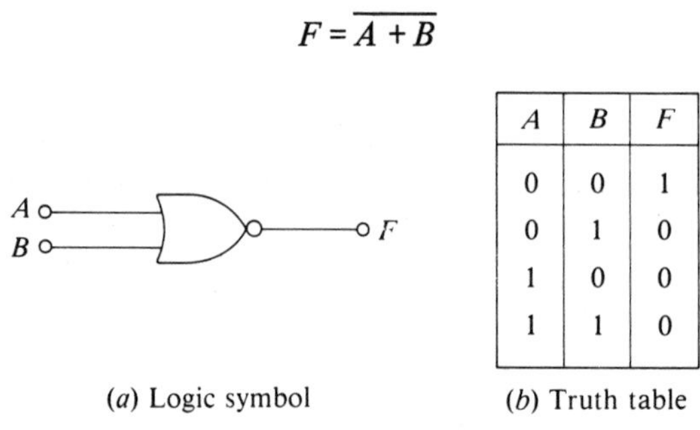

A	B	F
0	0	1
0	1	0
1	0	0
1	1	0

(*a*) Logic symbol (*b*) Truth table

Fig. 12.4 The NOR gate.

12.8 The NAND gate

This is NOT-AND, and may be realized using an AND gate followed by a NOT gate. The symbol and truth table for the NAND gate are shown in Fig. 12.5. The Boolean equation is

$$F = \overline{A.B}$$

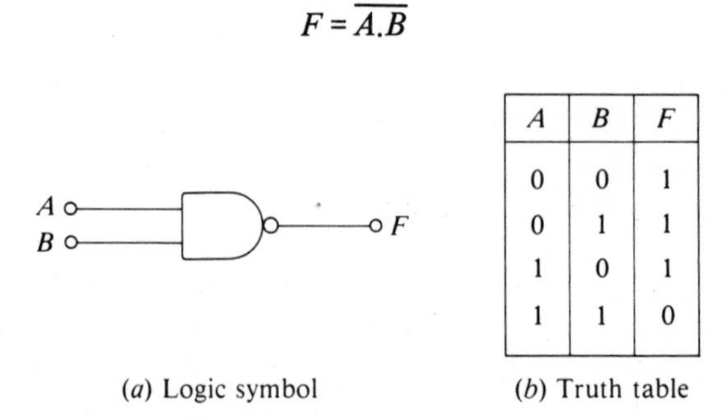

A	B	F
0	0	1
0	1	1
1	0	1
1	1	0

(*a*) Logic symbol (*b*) Truth table

Fig. 12.5 The NAND gate.

12.9 The laws of logic

Provided that the logical statement is accurate, it is possible to test its truth. Using binary notation, we say that if a statement is true, i.e., if the function exists, it has a logical value of 1. If it is false, it does not exist, and has a logical value of 0.

There are many laws of logic, but they are all derived from the following basic list:

1. $A.0 = 0$
2. $A + 0 = A$
3. $A.1 = A$
4. $A + 1 = 1$
5. $A.A = A$
6. $A + A = A$
7. $A.\overline{A} = 0$
8. $A + \overline{A} = 1$

These laws may be verified by using the simple switching arrangements shown in Fig. 12.6, in which AND functions are represented by series-connected switches and OR functions are represented by parallel-connected switches.

If a complemented signal is applied to a NOT gate, the output of the gate will be the original signal:

$$\overline{\overline{A}} = A$$

In addition to the laws listed above, two more are particularly useful when simplifying Boolean equations. These are DeMorgan's Dual:

$$\overline{A.B.C} = \overline{A} + \overline{B} + \overline{C}$$

and

$$\overline{A + B + C} = \overline{A} + \overline{B} + \overline{C}.$$

12.10 Choice of logic family

Electronic logic elements have evolved through a number of stages, beginning with systems consisting of *diode* AND and OR gates. Advances in semiconductor technology fostered rapid developments in electronic logic circuitry of the active type, and various circuits were produced. The first *integrated* logic elements were simply translations of discrete component circuits directly into silicon circuits. The earliest types were, in fact, composed of several silicon chips with wire interconnections. As integrated circuit techniques developed, the design approach changed and the circuits began to be designed to suit the manufacturing technology, instead of being duplicates of discrete component prototypes. Once it was realized that circuit complexity was not a limiting factor, the way was open for the production of high-performance, complex circuit elements.

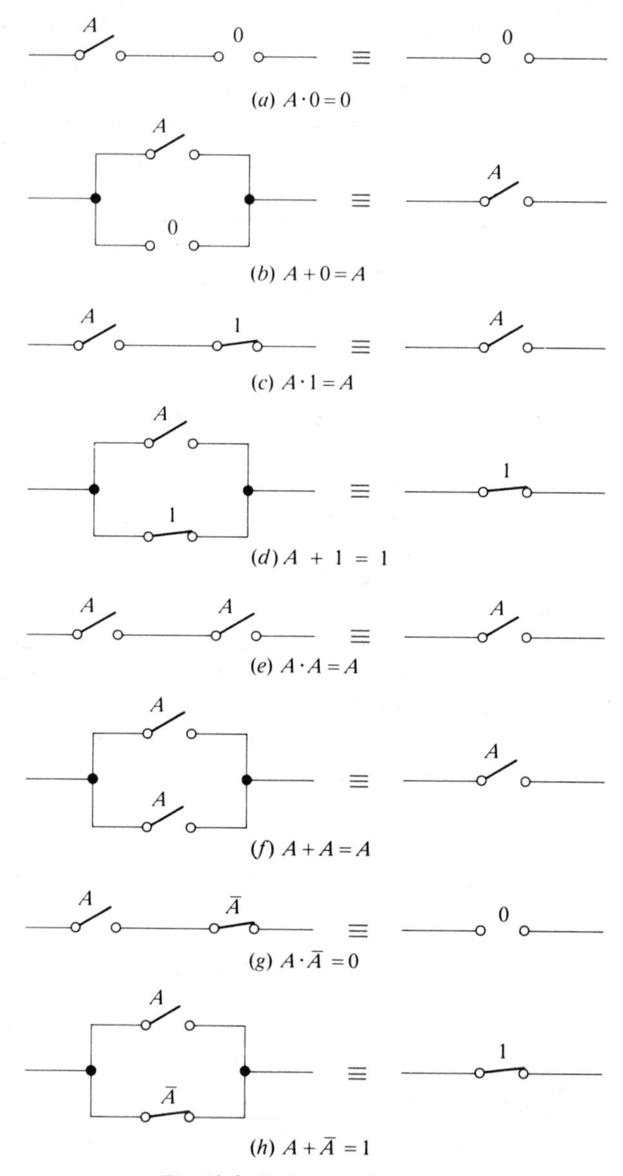

Fig. 12.6 Basic laws of logic.

Unfortunately, integrated circuit manufacturers tend to introduce their own designs of logic circuits with little standardization in mind. Eventually, however, certain types emerge as being more popular and these are duplicated by other manufacturers, thus providing ‘multiple source availability’.

It is not possible to draw the circuit diagram of a modern integrated circuit – even if we should want to do so. However, in order that we may use conventional theories to confirm functions for ourselves, the manufacturers include *functional* or *schematic* diagrams (*not* circuit diagrams). Unfortunately, we use these diagrams so much that we begin to believe that they are the *actual* circuits.

It has been common practice to classify logic families by the circuit configuration of the basic gate technology, the earlier types being ***D***iode ***R***esistor ***L***ogic (*DRL*), ***R***esistor ***T***ransistor ***L***ogic (*RTL*), ***D***iode ***T***ransistor ***L***ogic (*DTL*) which are briefly included here to demonstrate a number of important electronic principles.

12.11 Diode Resistor Logic (DRL)

1. ***Diode AND gate.*** The AND logic function may be realized with two diodes and a resistor arranged as shown in Fig. 12.7(*a*).

   If either input *A* or *B* or both are low, i.e., connected to 0 V (logical '0') then that diode is forward biased and current flows through the resistor *R* and the diode. The output voltage at *F* is then the voltage drop across the diode (0.7 V for silicon, 0.3 V for germanium), and this is the logical '0' output level.

   If both *A* AND *B* are high, i.e., connected to + V_{CC} (logical '1'), then both diodes are reverse biased, no current flows through the diodes and the output voltage is + V_{CC} (logical '1').
2. ***Diode OR gate.*** The OR logic function may be realized with two diodes and a resistor arranged as shown in Fig. 12.7(*b*).

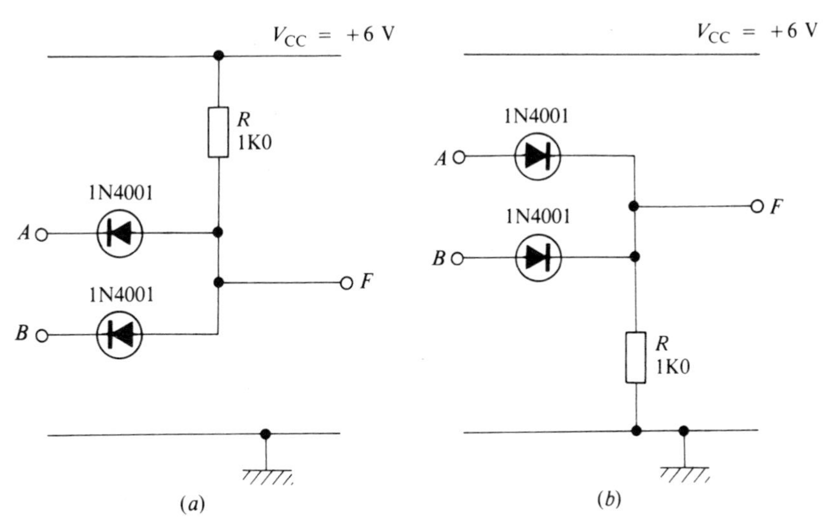

Fig. 12.7 Diode Resistor Logic (DRL) gates. (*a*) AND gate; (*b*) OR gate.

If either or both inputs A or B are at logical '1', the diodes are forward biased and current flows through R. The output voltage at F is, thus, $(V_{CC} - 0.7)$ V for silicon diodes, and $(V_{CC} - 0.3)$ V for germanium, i.e., logical '1'.

If both A and B are at logical '0', then no current flows through R and the output at F is 0 V, i.e., logical '0'.

12.12 Resistor Transistor Logic (RTL)

A simple RTL NOR gate can be realized with the circuit shown in Fig. 12.8.

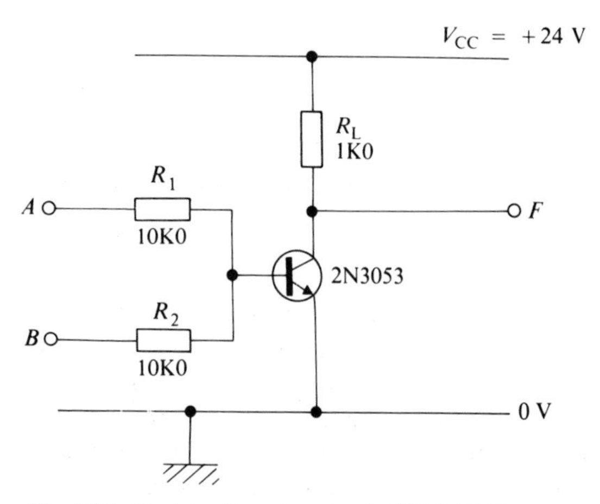

Fig. 12.8 Resistor Transistor Logic (RTL) NOR gate.

This arrangement uses the switching capabilities of the transistor. When inputs at A and B are low (logic '0') the transistor is OFF and the output voltage at F is high at + V_{CC}, i.e., logic '1'. When either A or B or both are high (logic '1') the transistor switches ON and conducts, and the output voltage at F falls to $V_{CEsat} \approx 0.2$ V for silicon, i.e., logic '0'.

12.13 Diode Transistor Logic (DTL)

A simple DTL NAND gate can be realized with the circuit shown in Fig. 12.9. This can be seen to consist basically of a diode AND gate followed by a transistor inverter.

Practical Exercise 12a

Diode logic gates

Connect up the diode AND gate shown in Fig. 12.7(a) and use the DMM to measure the output voltage at F for all the combinations of inputs. Logic '0'

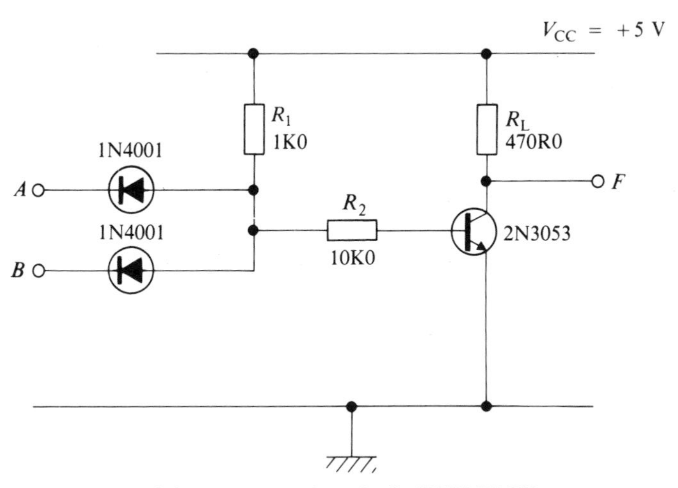

Fig. 12.9 Diode Transistor Logic (DTL) NAND gate.

input signals are achieved by connecting the input to 0 V. Logic '1' input signals are achieved by connecting the inputs to $+V_{CC}$. Hence draw up the truth table.

What happens if the output is loaded with a 6 V lamp?

Connect up the diode OR gate shown in Fig. 12.7(*b*) and measure the output voltage at F for all combinations of input signal. Hence draw up the truth table.

What happens if the output is loaded with a 6 V lamp?

If, *ideally*, logic '1' = + V_{CC} and logic '0' = 0 V, how do the two diode gates compare to the ideal? Which is the better gate?

Practical Exercise 12b

Resistor Transistor Logic

Connect up the RTL gate shown in Fig. 12.8 and use the DMM to measure the output voltage for all input combinations. Hence draw up the truth table.

Investigate the *quality* of the logic 1 and logic 0.

Measure the input current to cause the transistor to change its logical state with the component values shown.

Investigate the effect of loading the output with other similar gates. If it is assumed that the minimum high level at the output is 18 V, determine the *fan-out* of this gate.

Fan-out is the maximum number of gates that may be driven whilst maintaining its output voltage within specified limits.

How could this value of fan-out be increased?

Note: With 10K0 input resistors, the drive current is clearly more than enough to saturate the transistor ($V_{CEsat} \approx 0.2$ V).

When driving a similar gate, the output feeds into a 10K0 input resistor in series with the base–emitter junction of the driven transistor (about 1 kΩ). Neglecting the transistor input impedance, this represents a load of about 10K0 connected across the output of the driving gate. Further driven gates may be represented by additional 10K0 resistors in parallel with the first. Each 10K0 'load' carries approximately the same value of the drive current, which must come from the supply – and hence flow through the 1K0 collector load. To maintain our output voltage level above 18 V, this means that the collector current must be limited to 6 mA (i.e., 6 mA x 1K0 = 6 V), thus restricting fan-out to about 3.

If the 10K0 input resistor is connected in series with a 1M0 variable resistor, and the base input (drive) current measured while the collector–emitter voltage is monitored, then the minimum drive current may be determined to saturate the transistor. If each RTL gate uses the increased base input resistance then the fan-out will be increased.

Practical Exercise 12c

Diode Transistor Logic

Connect up the DTL gate shown in Fig. 12.9, and use the DMM to measure the output voltage for all combinations of input signal. Hence draw up the truth table.

12.14 IC logic families

The early types have been largely superseded by the following IC logic families:

Transistor ***T***ransistor ***L***ogic (TTL)
Emitter ***C***oupled ***L***ogic (ECL)
Complementary ***M***etal ***O***xide ***S***ilicon Logic (CMOS)

TTL and CMOS have evolved as the most popular, but use different technologies in their manufacture (TTL–*bipolar*, CMOS–*unipolar*) which have different handling requirements.

When handling CMOS devices great care must be taken since they are very susceptible to damage from excessive voltage caused by static electricity and equipment which is not correctly earthed. These devices are usually supplied with their pins embedded in conductive foam; KEEP THIS INTACT UNTIL THE CIRCUIT IS TO BE CONNECTED. *DO NOT USE PLASTIC OR NYLON MATERIAL*, since these generate high static voltages. Similarly, take care with clothing – many of the man-made fibres used today also generate large amounts of static electricity. In addition, all equipment used must be properly earthed. Working with CMOS has therefore created new problems, most of which can be overcome without too much difficulty. One method which has been used very successfully is that of a copper plate working surface properly earthed together with all other equipment – even to the extent of conductive 'wrist-bands' connected to the same earth. However, most CMOS ICs available today have some internal protection (buffered inputs) but, even with these it is important to connect the supply pins before the protective circuitry can become effective. Finally, DO NOT LEAVE UNUSED INPUTS FLOATING, always connect unused inputs – either to used inputs or to the appropriate supply rail for the logic used. Input signals must not be applied until the power supply is connected, and is ON.

Some of the disadvantages of CMOS ICs have been emphasized here in an attempt to stress the importance of *'handle with care'*, since many heartaches have been caused by not observing simple precautions.

The main advantages of CMOS are high packing density, i.e., more complex circuits are possible on a given size of silicon chip, wide range of supply voltage (3 V to 15 V), high noise immunity (the degree to which a gate can withstand variations in input levels without causing a significant change in output state), and very low power consumption. Operating from a 5 V supply, typical values are: propagation delay 35 ns, power dissipation 10 nW, and noise immunity 2 V.

12.15 Working with IC logic devices

In practice, logic systems are designed and constructed using either NAND gates only or NOR gates only. Now, the only logical *functions* that need to be performed are AND, OR, and NOT, and these functions may be realized using NAND gates only *and* using NOR gates only.

Using NAND logic, unused inputs must *either* be connected to a used input *or* to a logical HI level (logic 1), i.e., supply rail through a 1K0 resistor (220K0 for CMOS).

Using NOR logic, unused inputs must *either* be connected to a used input *or* to a logical LO level (logic 0), i.e., supply ground.

Therefore, the NOT logical function can be performed as shown in Fig. 12.10.

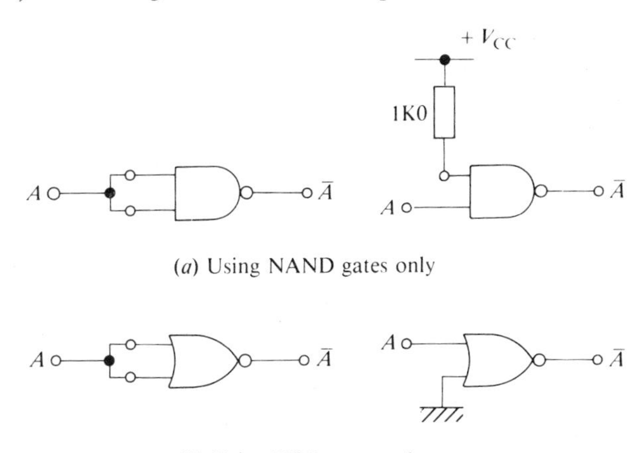

Fig. 12.10 The NOT function using NAND and NOR gates.

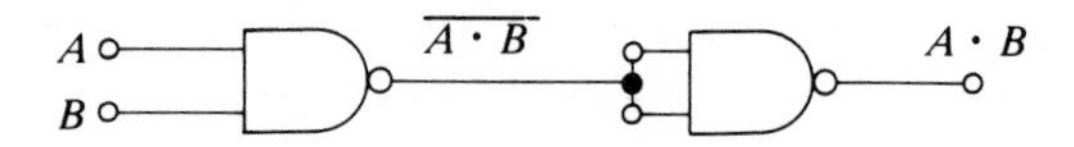

(*a*) The AND function

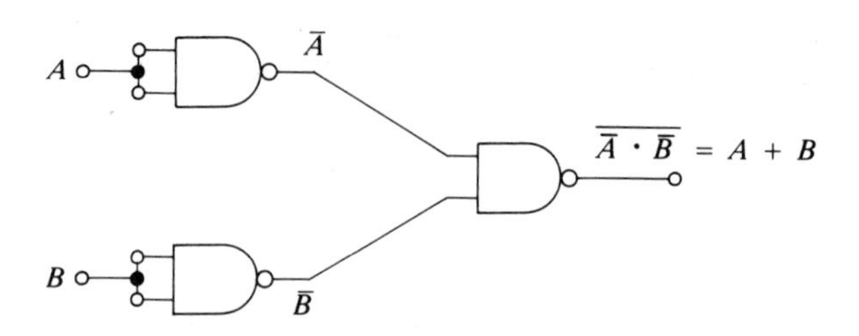

(*b*) The OR function

Fig. 12.11 The AND and OR functions using NAND gates only.

The AND and OR functions may now be performed using NAND gates only as shown in Fig. 12.11, and using NOR gates only as shown in Fig. 12.12. It should be noted, however, that whichever logic family is used, NOT gates are usually available – frequently six in a package – and are then referred to as *Hex Inverters.*

Avoid having 'floating' inputs. This is particularly important for those logic

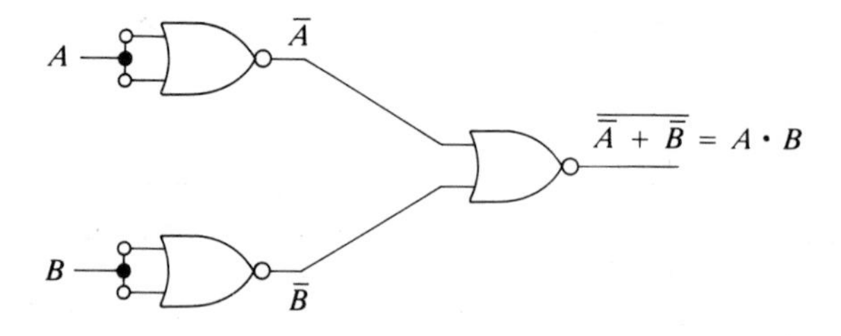

(*a*) The AND function

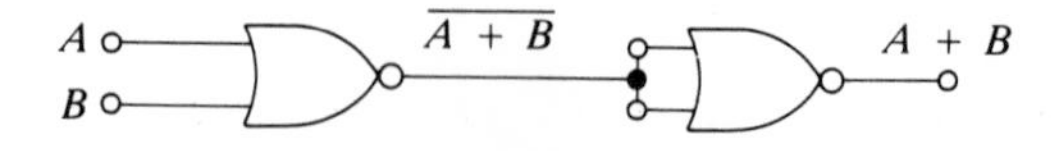

(*b*) The OR function

Fig. 12.12 The AND and OR functions using NOR gates only.

devices which respond at the leading or trailing *edge* of a *changing* logic state. Therefore, it is important to use *debounced* switches for selecting logic levels or states.

12.16 Transistor Transistor Logic (TTL)

This logic family has been by far the most popular until recently, and may be considered as a development of DTL, in which the input diodes are replaced by the emitter–base junctions of a multi–emitter transistor (easily fabricated using the advanced technology), as shown in the functional diagram in Fig. 12.13.

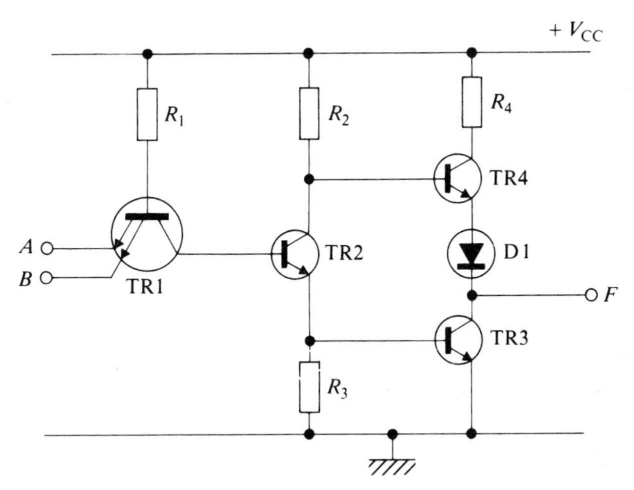

Fig.12.13 Functional diagram of TTL NAND gate.

When both inputs at *A* and *B* are at logic 1, the emitter–base junctions of TR1 are reverse biased, and sufficient current flows through R_1 and the base–collector junction of TR1 to provide base drive to switch TR2 hard on, so that it holds TR3 in saturation and maintains TR4 at cut-off, and the output at *F* is logical 0.

When any or all of the inputs are at logic 0, current flows *out* of the corresponding emitter of TR1. This removes the base drive from TR2, causing it to be cut-off, which in turn removes the drive from TR3 – causing it to be cut-off. Current now flows through R_2 to drive TR4 into saturation, and the output at *F* is logical 1.

The push-pull nature of the output stage is referred to as a *totem pole* arrangement – only *one* of the transistors TR3 or TR4 can be on at any instant.

TTL has become a particularly popular range, and has been around for over 20 years. The standard (*normal*) TTL gates have typical figures of 10 mW power dissipation, 10 ns propagation delay, and about 0.5 V noise immunity. The popular SN7400 series are widely available from many sources, and the TTL logic levels are:

Logic 1 = greater than 2.4 V (typically about 3.6 V).
Logic 0 = less than 0.4 V (typically about 0.2 V).

The input level at which the gates change their state is between 0.8 V and 2.0 V.

Practical Exercise 12d

TTL NAND gates

Consult manufacturer's data for the pin connections (pin-outs) for the 74 series TTL gates.
Select the SN7400 Quad 2 input NAND gate – as shown in Fig. 12.14. The supply voltage for these TTL gates is 5 V ± 0.25 V.

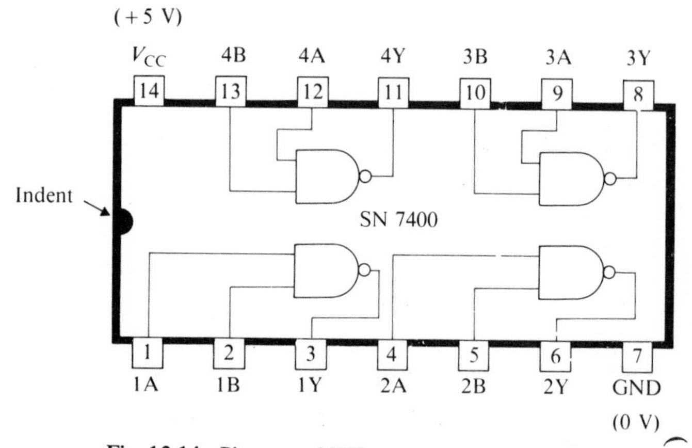

Fig. 12.14 Pin-outs of TTL NAND gate (SN 7400).

Apply all the combinations of input logic levels to the inputs of one gate and monitor the logic state of the output. Thus, construct the truth table for the NAND gate.

Repeat this exercise, but measure the *actual* voltages at the output and compare with the values stated above.

Interconnect NAND gates as shown in Fig. 12.11, and verify the operation of the AND and OR functions.

12.17 Sinking and sourcing currents

When one TTL gate drives another TTL gate, the limiting currents are specified in the manfacturer's data.

1. ***Sink current.*** When the output of the *driving* gate is low, i.e., logical 0 (or LO) level, the path for the current flow is as shown in Fig. 12.15(*a*), and the driving gate output (TR3) is said to *sink* the current, i.e., provide a current path to earth. The specification for standard TTL (having a nominal fan-out of 10) states that the maximum sink current I_{OL} is 16 mA. Therefore, the maximum sink current for each driven gate is 1.6 mA. (Typical value 1.1 mA).

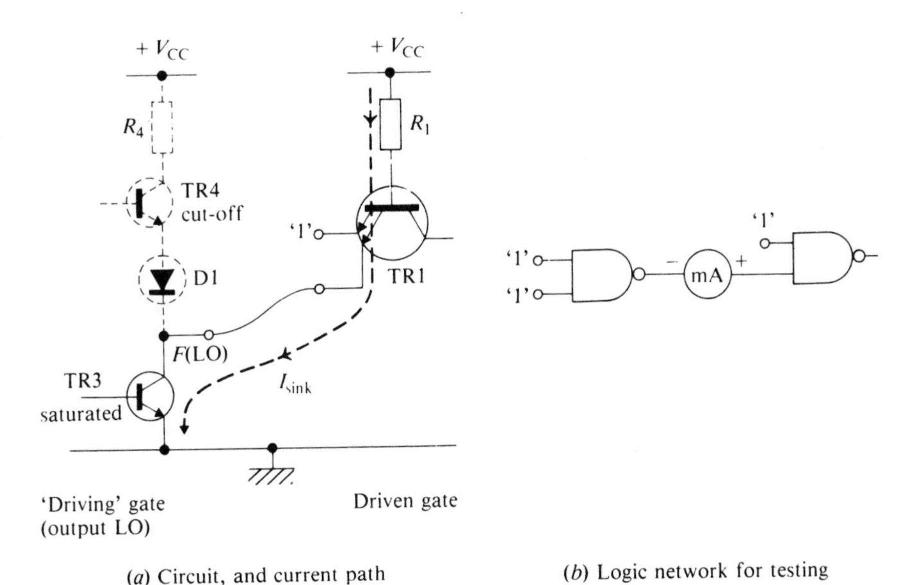

(*a*) Circuit, and current path

(*b*) Logic network for testing

Fig. 12.15 TTL sink current.

Practical Exercise 12e

TTL sink current

Connect up two TTL NAND gates, with 0–10 mA milliammeter, as shown in Fig.12.15(*b*). *Note the polarity of the instrument.* Before switching on the supply, ensure that the logic levels will be as shown at switch-on.

Switch on the supply and record the instrument reading. If this is less than 1.6 mA, it may be assumed to be satisfactory. Switch off the supply.

Note: Always switch off the supply, whenever making any changes to interconnections on logic chips.

2. ***Source Current.*** When the output of the *driving* gate is high, i.e., logical 1 (or HI) level, the path for the current flow is as shown in Fig. 12.16(*a*), and the driving gate output (TR4) is said to *source* the current drive for the driven gate. The manufacturer's specification states that the source current I_{OH} is a maximum of 400 μA (for a nominal fan-out of 10). Therefore, the maximum source current for each output is 40 μA. (Typical value 5 μA).

Practical Exercise 12f

TTL source current

Connect up two TTL NAND gates, with a 0–100 μA microammeter, as shown in Fig. 12.16(*b*) *Note the polarity of the instrument.*

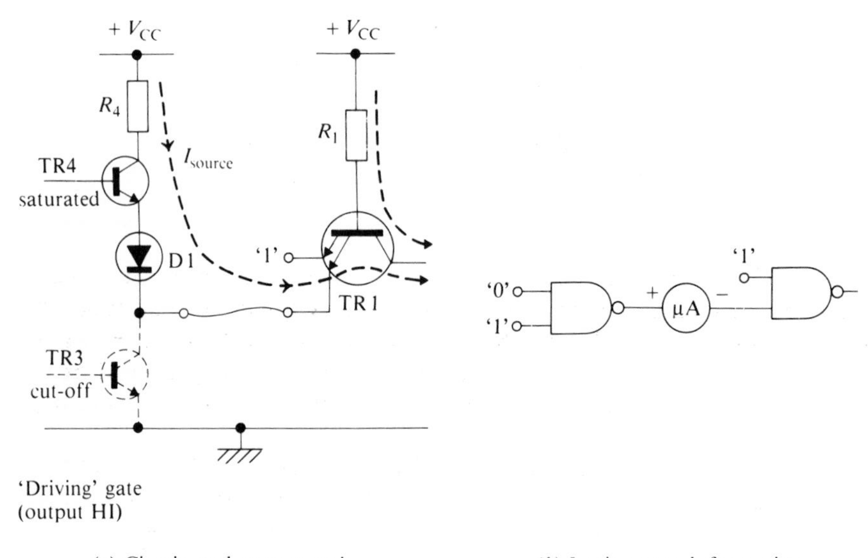

(*a*) Circuit, and current path (*b*) Logic network for testing

Fig. 12.16 TTL source current.

Before switching on the supply, ensure that the logic levels will be as shown at switch-on.

Switch on the supply and record the instrument reading. If this is less than 40 μA, it may be assumed that the gate is satisfactory. Switch off the supply.

Practical Exercise 12g

Exclusive-OR function and comparator

Connect up the logic network shown in Fig. 12.17(*a*) using *one* SN7400 Quad

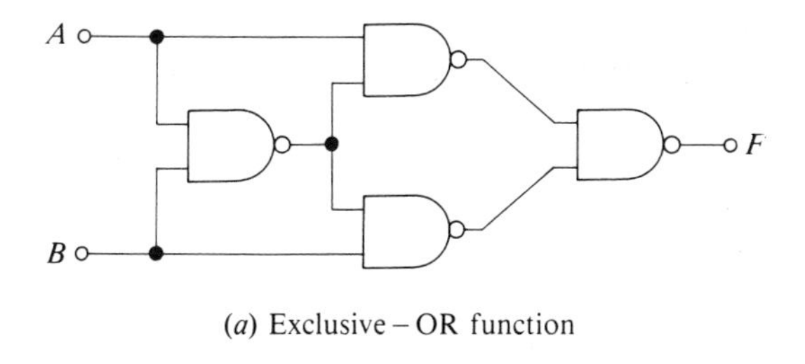

(a) Exclusive – OR function

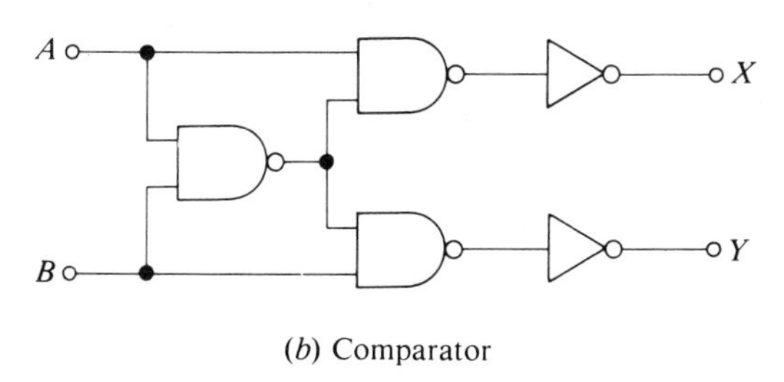

(b) Comparator

Fig. 12.17 Exclusive-OR function and comparator.

2 input NAND gate and LEDs as logic state indicators. Apply the various combinations of inputs (from 'binary' switches, switching between logic '0' = 0 V and logic '1' = + 5 V) and check the output states with the LEDs. Hence draw up the truth table. This is an exclusive-OR logic function, giving an output '1' when the two inputs are different, and an output '0' when the two inputs are the same.

Modify the circuit to that shown in Fig. 12.17(b), using 1/3 SN 7404 – Hex Inverter. Connect the outputs X and Y to two LEDs. Hence, determine the combinations of input signals for which the outputs at X and Y are logic '1'.

Which gives an output when A is greater then B, i.e., $A = 1, B = 0$?

Which gives an output when A is less than B, i.e., $A = 0, B = 1$?

Thus, this simple network is capable of '*comparing*' two binary signals, and is called a *comparator.*

12.18 Counting systems

One of the most important types of electronic digital system is the digital computer and its associated control circuits. The digital computer is essentially a calculating machine and although the basic principles of its operation are very simple, it is because the computer is capable of solving complex mathematical problems very very quickly that it has the overwhelming aura of complexity.

Many industrial processes use a computer to 'control' the process. In this respect, the computer performs the *comparison* calculation and produces an answer which is converted to a suitable form to apply to the process. Devices and equipment forming the interconnection between process and computer is termed *interface* equipment.

Many other applications of digital systems (i.e., systems using digital logic) exist in all aspects of everyday life, e.g., pelican crossing, traffic lights, lift operation and control. All of these systems depend on sequential and combinational logic systems, together with a certain amount of timing circuitry, which implies the need for the system to be able to count pulses for a given period.

12.14 Numbering systems

1. ***Denary system.*** This is the system we are all very familiar with. The denary system uses *ten* symbols representing the quantities '0' through to '9'. The number of symbols used in the system is called the ***base***, or ***radix***, i.e., ***ten*** in this case.

   Other numbers are constructed by giving different values, or *weights*, to the position of the digit relative to the 'decimal' point. The weights of the different positions are given by powers of the radix, which, in general, is

$$R^2 \; R^1 \; R^0 \cdot R^{-1} \; R^{-2}$$

   and, for the denary system, is

10^2	10^1	10^0	.	10^{-1}	10^{-2}
Hundreds	Tens	Units	.	Tenths	Hundredths

Example

$426_{10} = 4 \times 10^2 + 2 \times 10^1 + 6 \times 10^0$

$= 400 + 20 + 6$

$= 426.$

   We are so familiar with this system that we tend not to think about the way in which the numbers are constructed.

2. ***Octal system.*** This was widely used in computer programming. The ***radix*** is ***eight***, and positional weights are powers of eight.

Example

$426_8 = 4 \times 8^2 + 2 \times 8^1 + 6 \times 8^0$

$= 4 \times 64 + 2 \times 8 + 6 \times 1$

$= 256 + 16 + 6$

$= 278_{10}$

3. ***Binary system.*** This is widely used in logical systems and computing. The *radix* is *two*, and positional weights are powers of two:

$$2^3 \; 2^2 \; 2^1 \; 2^0 \; . \; 2^{-1} \; 2^{-2}$$

Since the radix is *two*, only two digits, 0 and 1 exist, so that in this system 1 + 1 = 0 carry 1, i.e., 10, one-nought, *not* ten.

A table of comparisons of number systems is shown in Fig. 12.18.

Denary		Octal		Binary			
Tens 10^1	Ones 10^0	Eights 8^1	Ones 8^0	Eights 2^3	Fours 2^2	Twos 2^1	Ones 2^0
	0		0				0
	1		1				1
	2		2			1	0
	3		3			1	1
	4		4		1	0	0
	5		5		1	0	1
	6		6		1	1	0
	7		7		1	1	1
	8	1	0	1	0	0	0
	9	1	1	1	0	0	1
1	0	1	2	1	0	1	0
1	1	1	3	1	0	1	1
1	2	1	4	1	1	0	0

Fig 12.18 Comparison of number systems.

12.20 Binary coded decimal (BCD) systems

Information or data is conveyed in logic systems and computers in binary form, since this is more convenient for logic devices (although it is not very suitable for humans). By *coding* decimal numbers in binary form, suitable numbering systems are devised which are acceptable to both the human *and* the machine.

Any decimal number can be represented by a *group of four binary digits.* An example of this is the 8421 BCD as shown in Fig. 12.19.

After a count of 9_{10}, the 8421 BCD weights change by a factor of ten. Thus, the next four *bits* are 80, 40, 20, 10, enabling a decimal number up to 99 to be represented.

12.21 The half-adder

Basic logic networks may be used in counting systems. The simple exclusive-OR logic function is effectively a *half-adder*, giving an output '1' when $A = 1, B = 0$

Denary	8421 BCD
0	0000
1	0001
2	0010
3	0011
4	0100
5	0101
6	0110
7	0111
8	1000
9	1001

Fig. 12.19 8421 BCD system.

and when $A = 0$, $B = 1$, and giving an output '0' and a carry digit '1' when $A = 1$, $B = 1$, as shown in Fig. 12.20

In practical calculations, it is necessary to account for the carry digit generated by the previous calculation. Thus, several half-adders may be interconnected to form a binary counter.

Counting systems usually use bistable (incorrectly called *flip-flops*) devices, which have two stable states, in which the application of a signal causes the device to change from one stable operating state to the other.

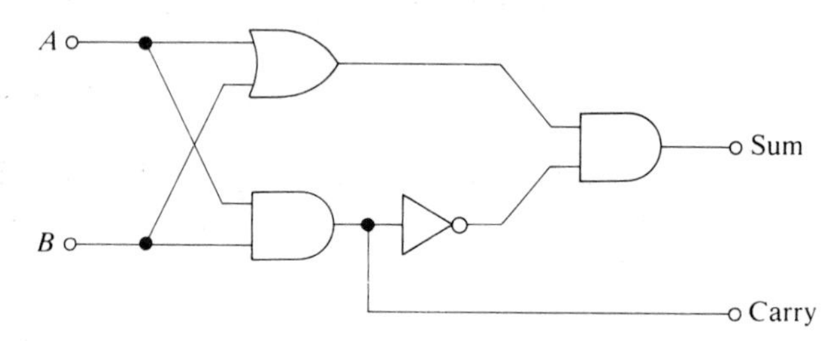

Fig. 12.20 Exclusive-OR network as half-adder.

12.22 The S–R flip-flop (set-reset)

The symbol for the S–R flip-flop is shown in Fig. 12.21(*a*) and a logic network using NOR gates only, capable of performing the same function is shown in Fig. 12.2(*b*).

A logical '1' signal applied to S causes the output Q to be set to '1' irrespective of its previous state (at the same time, the output $\overline{Q}$ becomes '0'). A logical '1' signal applied to R causes the output Q to be *reset* to '0'.

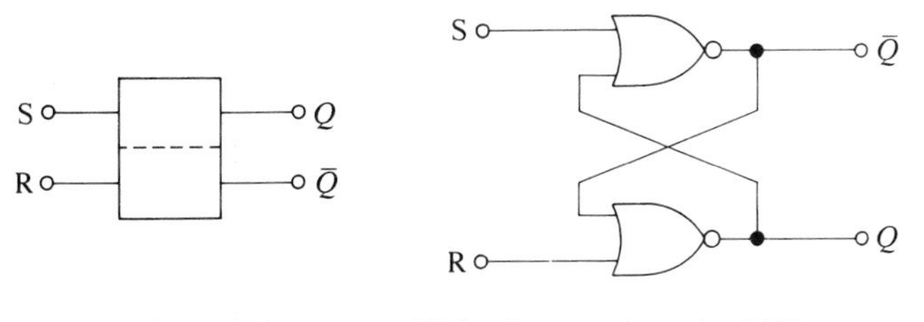

(*a*) Logic symbol (*b*) Logic network – using NOR gates only

Fig. 12.21 The S–R 'flip-flop'.

Logical '1' signals simultaneously applied to S and R result in an indeterminate output stage – a condition which must be avoided. However, simultaneous application of logical '1' signals to the S and R inputs of the cross-coupled NOR network causes *both* outputs to be logical '0'.

12.23 Clocked J–K flip-flop

In addition to overcoming the limitation of the S–R flip-flop, this flip-flop has a clock input terminal, so that if J and K are connected together and to logic '1', the flip-flop will change its state each time a clock pulse is applied to the clock input terminal.

12.24 Simple binary counter

The simple binary (or serial) counter is an asynchronous counter in which the pulses to be counted are applied at one end of the counter, and the process of adding each pulse must be completed before the 'carry bit' is propagated to the following stage. This next stage must then add the carry bit to the number in that stage. Thus, the carry bit appears to *ripple-through* the length of the counter until the count is complete. These systems are often referred to as *ripple-through counters*, as shown in Fig. 12.22.

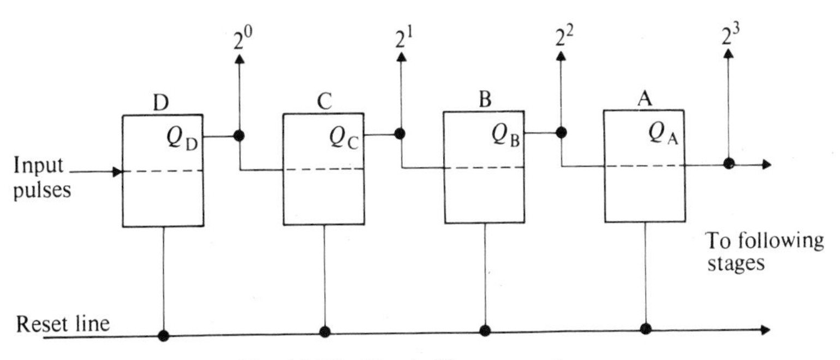

Fig. 12.22 Simple binary counter.

12.25 The decade counter

The simple binary counter shown above may be converted into a decade counter by using a simple logic network *decoder* to detect the state 9_{10}, i.e., binary 1001 and apply a logic '1' to the reset line to reset all flip-flops to '0', and re-start the count. Owing to the switching characteristics of the flip-flops, it is necessary to detect the following state to that required. Thus, in this case, binary 1010 is the state to be detected, which may be achieved with a two-input AND gate (Q_A and Q_C being applied to the inputs of the AND gate), and its output connected to the 'reset to 0' line.

12.26 BCD counters

BCD counters may be constructed by similar techniques to those suggested above.

Similarly, many industrial processes use *batch* counters to count a desired number of components, e.g. testing, sorting, packaging.

Practical Exercise 12h

8421 BCD counter and decade counter

Connect up the arrangement shown in Fig. 12.23 and observe the output states of the flip-flops (within the SN 7490) on the LEDs as a train of pulses is applied to the input. Note the effect of switching the *reset* to 0 (R_0) signal to logical '1' and back again to logical '0'.

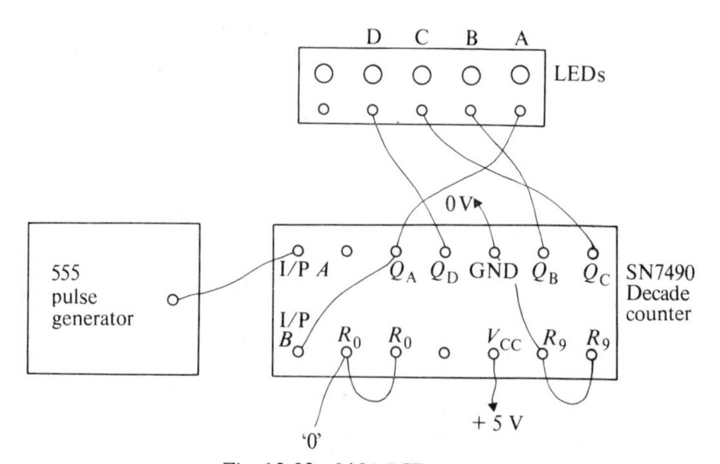

Fig. 12.23 8421 BCD counter.

Now connect up the decade counter arrangement, using the decoder/seven-segment driver and seven-segment display as shown in Fig. 12.24, and observe

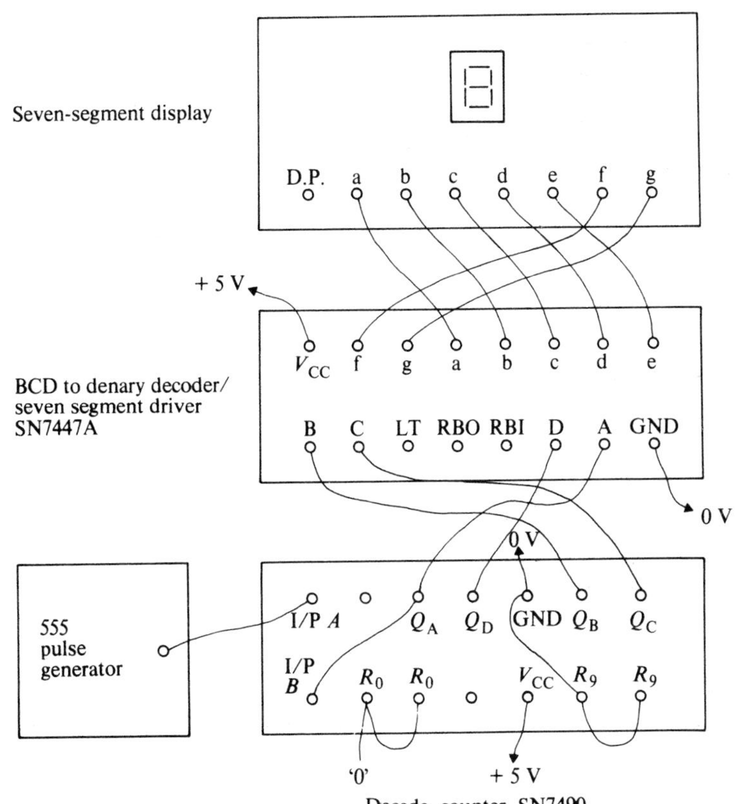

Fig. 12.24 Decade counter.

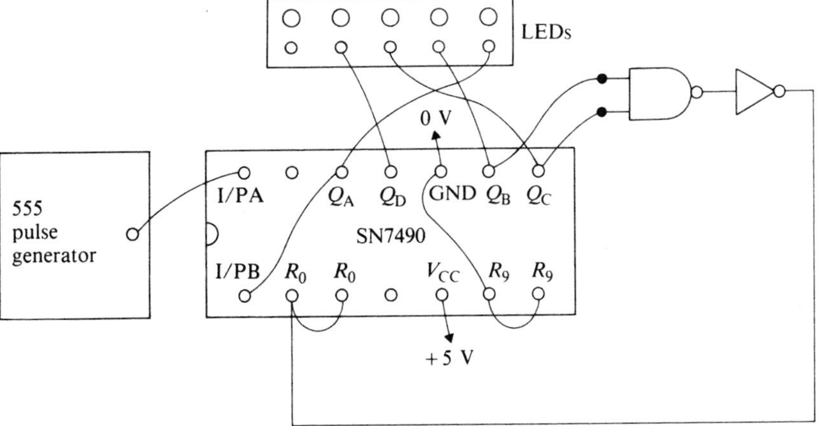

Fig. 12.25 Batch counter.

the display as a train of pulses is applied to the input. Note the effect on the display of switching the R_0 signal to logical '1' and back to '0'.

Practical Exercise 12i

Batch counter

Connect up the arrangement shown in Fig. 12.25. Observe the output display as a train of pulses is applied, and determine the number in the batch.

Investigate various modifications to the decoding network to one capable of counting a batch containing a different number.

13. Industrial Control

13.1 Introduction

The field of industrial control engineering is vast; it includes all branches of science and utilizes any form of energy, e.g., mechanical, electrical/electronic, hydraulic, pneumatic. Control systems have been devised for a wide range of industrial applications, e.g., voltage, power, temperature, level, flow, pressure, position, velocity (speed), etc.

However, whichever form of energy is used, and whichever physical variable is being controlled, the basic principles of control are essentially the same. Many industrial processes now use electronics to produce and 'condition' the required controlling signal due to the advantages of physical size, reliability, and – with the rapid introduction of microelectronics – cost.

13.2 Open-loop control

This is the simplest form of control; it is inexpensive and may be adequate in many cases. In open-loop control systems, an action is initiated but there is no way for the system to modify that action as it responds, i.e., there is no *feedback*, as shown in Fig. 13.1. The domestic light circuit is a common example of open-loop control, in which the initiating action is the operation of the switch – there is no means by which the position of the switch can be changed if the light intensity is insufficient, or too much.

Open-loop systems are characterized by systems using switches, contactors, valves, etc. In many cases, a human operator may observe the prevailing state of the controlled variable. In such cases, if the human operator takes any action – such as adjustment of a controller – then the human operator acts as a *feedback* path and the system becomes a *closed-loop* system for as long as the operator remains on duty.

13.3 Closed-loop control

As suggested above, the open-loop control system can be converted into a closed-loop system with a human operator acting as a feedback path, and taking actions

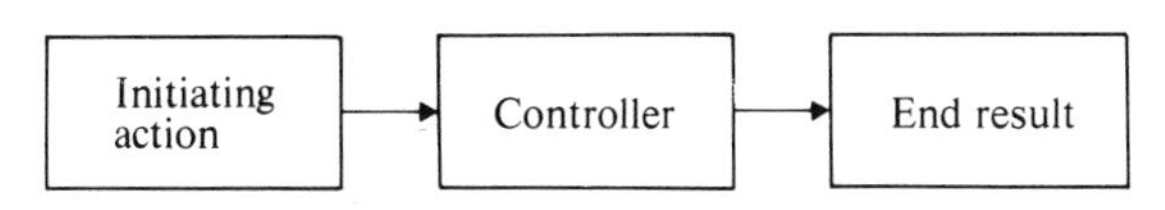

Fig. 13.1 Open-loop control system.

to maintain the controlled variable at a desired condition. In automatic closed-loop control systems, the object is to replace and improve the actions of the human operator in this role. This is generally achieved by making a measurement of (monitoring) the prevailing state of the controlled condition and converting it into a similar form to the initiating action (input signal); these two 'signals' are *compared*, and the difference (*error*) is the signal which is then used to control the system, i.e., automatic closed-loop control systems are error-actuated.

The monitoring and signal conversion may be generally achieved with one device – a *transducer*, which is a device capable of converting one kind of physical variable into another kind, e.g., an electrical transducer is one which converts a physical variable (velocity, temperature, etc.) into an electrical signal.

The error signal formed by the comparison of the input signal and the monitoring feedback signal may be rather small so that an amplifier may need to be used to increase the controlling signal. The block diagram of a closed-loop control system is shown in Fig. 13.2.

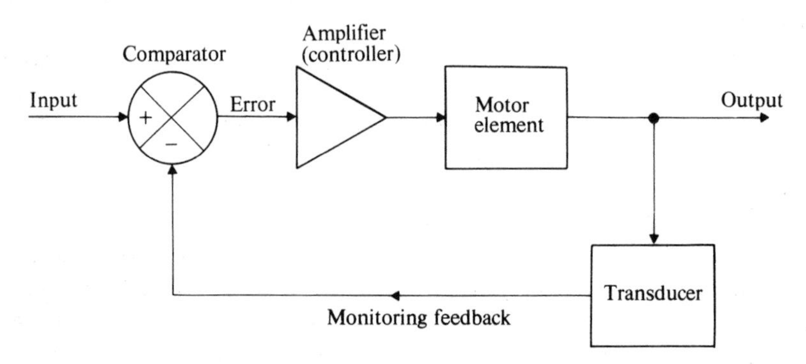

Fig. 13.2 Closed-loop control system.

13.4 Transducers

Many modern control systems use electrical transducers to produce the monitoring feedback signal. Electrical transducers have been made and are commonly available to deal with virtually all types of physical variable, some of the most common ones are as follows:

1. ***Load.*** Strain, force, pressure. Strain gauge – resistance changes $R = \rho l/a$.
2. ***Position.*** Displacement. Variable resistance (potentiometer), $R \propto l$. Linear variable differential transformer (LVDT). Capacitance, $C = \epsilon a/d$.
3. ***Velocity.*** Tachogenerator – d.c. or a.c. generator, optical.
4. ***Temperature.*** Thermocouple – e.m.f. proportional to change in temperature. Thermistor – resistance depends on temperature.

13.5 Types of control action

In a simple closed-loop control system with monitoring feedback, the controlling signal is *proportional* to the error, so that this is referred to as a *proportional control system*.

However, in most systems the presence of time lags does not allow the system to reset the controlled variable to the desired value after a disturbance occurs, but leaves an *offset* (or *droop*). This can be overcome by advancing the control signal to overcome the system lags – this action being known as *reset* or *integral control.* The control system is now referred to as a *two-term* control system, i.e., $P + I$ (proportional + integral).

Proportional control inherently produces some degree of oscillation of the controlled variable before finally settling to within acceptable limits of the desired value. The time taken to reach the desired value (settling time) can be decreased if we produce a control signal proportional to the *rate* of change of error. This is known as *rate* or *derivative* control and the control system is again referred to as a *two-term* control system, i.e., $P + D$ in this case (proportional + derivative).

The ultimate in response can be achieved by combining the above control actions to produce a *three-term* control system, i.e., $P + I + D$. The integral control action reduces the offset (or steady-state error) and the derivative control action reduces the oscillations, and improves the response time.

13.6 Power control

Since the early work of Heaviside and Steinmetz, electrical engineers have been deeply involved in the control of power. The thyristor has fulfilled the predictions that it would largely revolutionize the field of power control and inversion. More recently, the triac – replacing two thyristors connected in inverse-parallel – enables large amounts of a.c. power to be controlled at very high efficiencies.

13.7 Controlled rectification

We have previously considered (Chapter 5) the production of a d.c. voltage from an a.c. supply by various rectifier arrangements. The single-phase controllable rectifier consists of a conventional full-wave bridge rectifier arrangement in which two of the diodes are replaced by thyristors, as shown in Fig. 13.3(*a*). Each thyristor is pulsed once in each supply cycle, and continuous control of the d.c. output voltage is achieved by shifting the phase of the trigger pulses between 0° and 180° relative to the supply voltage half-cycle. Typical bridge output voltage waveforms with resistive loads are as shown in Fig. 13.3(*b*), and with inductive load as shown in Fig. 13.3(*c*).

13.8 Flywheel diode

When the thyristor is used to switch inductive loads, the current continues to

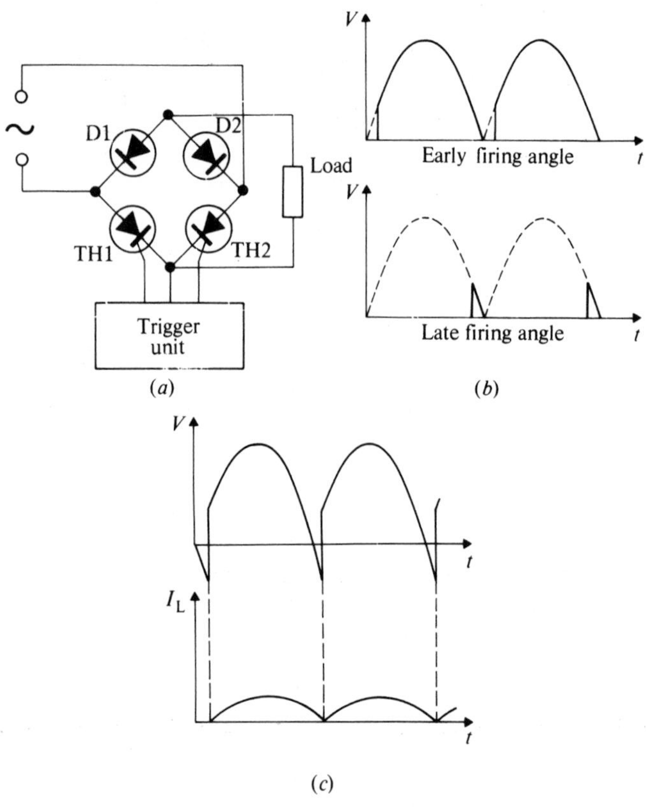

Fig. 13.3 Controllable bridge rectifier. (*a*) Circuit; (*b*) output voltage (resistive load); (*c*) waveforms with inductive load.

flow in the load when the thyristor has been switched off as shown in Fig. 13.3(*c*). The reverse current flowing in the load when the thyristor is switched off may be eliminated by using a *by-pass* or *flywheel* diode as shown in Fig. 13.4, which allows the current to circulate around the loop formed by the load and the diode.

Practical Exercise 13a

Full-wave controlled d.c. load

Connect up the circuit shown in Fig. 13.5, in which the trigger circuit is a standard UJT relaxation oscillator which derives its driving voltage from the same a.c. supply as the load. In this way the trigger pulses are synchronized to the a.c. power supply.

1. Apply the trigger pulses from the UJT trigger to both thyristor gates.

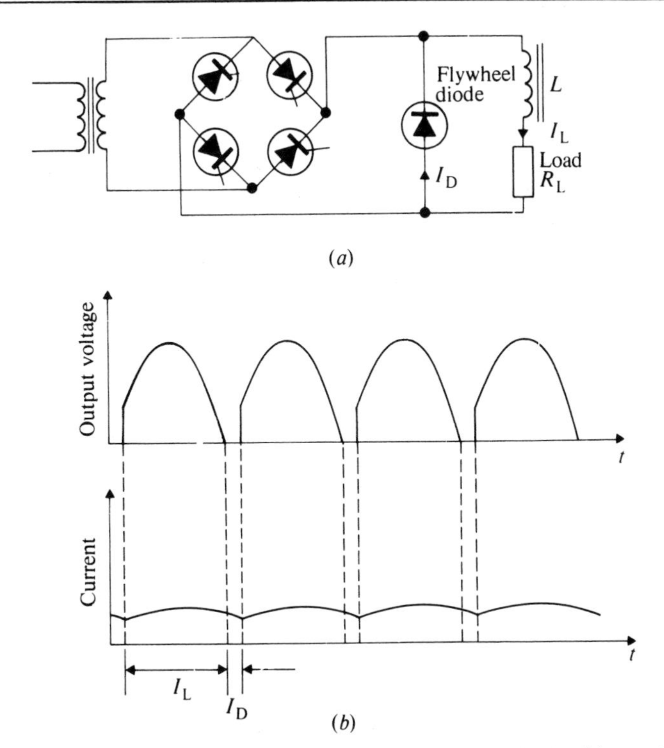

Fig. 13.4 Full-wave controlled bridge with flywheel diode. (*a*) Circuit; (*b*) waveforms.

2. Examine the waveforms across each thyristor and the load using a dual-trace CRO.
3. Vary the firing angle by adjusting the 25K0 variable resistor and observe the effect on the waveforms.
4. Remove the UJT trigger circuit from the thyristor gates and apply a trigger signal from the variable output of the function generator set to TRIANGULAR WAVE, 5 V peak-to-peak, and frequency range 1 – 200 Hz.
5. Sweep the frequency dial from end to end and note the effect on the lamp, and on the waveforms across the lamp and thyristors.

13.9 Bridge converter

A conventional full-wave rectifier arrangement in which all the diodes are replaced by thyristors is capable of both controlled rectification from a.c. to d.c., and of inversion from d.c. to a.c. This type of circuit is, therefore, particularly suited to regenerative loads, e.g., regenerative braking and reversing drives for d.c. motors.

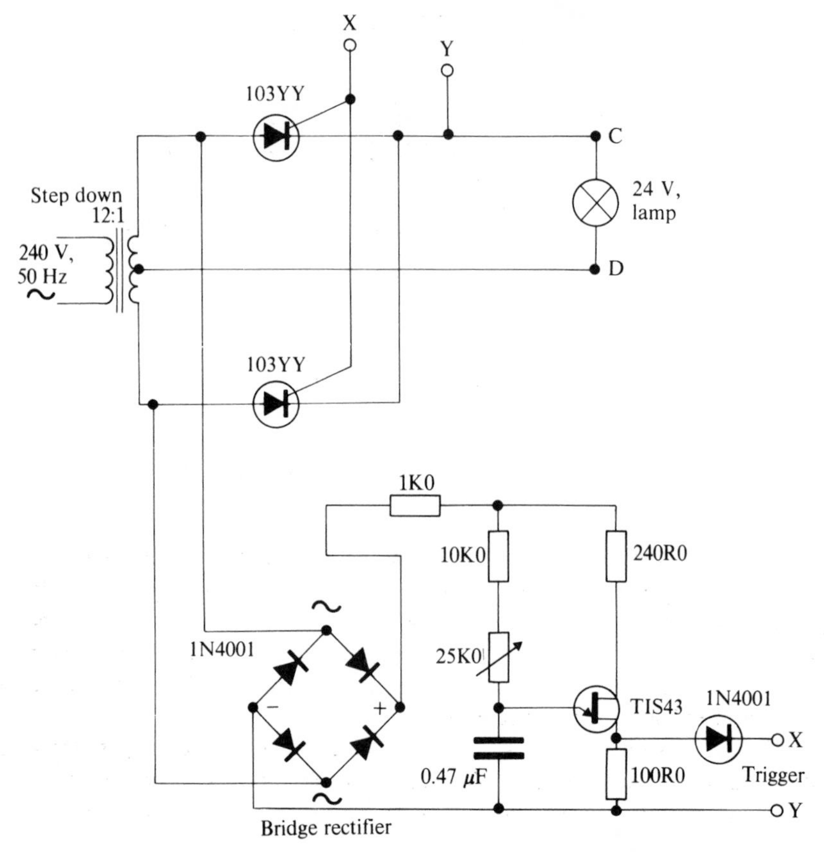

Fig. 13.5 Full-wave controlled d.c. load.

Controlled rectification is obtained in the normal way by shifting the phase of the driving pulses. In this case, opposite pairs of thyristors must be pulsed simultaneously.

For inversion to take place, it is necessary for a source of power to be connected in the d.c. side of the circuit. Since current flow at the d.c. terminals is inherently uni-directional, the polarity of the d.c. source must be connected in series aiding the rectifier polarity. The basic circuit of the converter is shown in Fig. 13.6.

The inverting action relies for its operation upon the presence of an a.c. supply, and in this respect the principle of operation is different from that of a free running inverter, which generates an alternating voltage from a direct voltage source.

Pairs of thyristors are simultaneously fired at a point in the supply cycle such that the a.c. voltage is connected in series opposition with, and is instan-

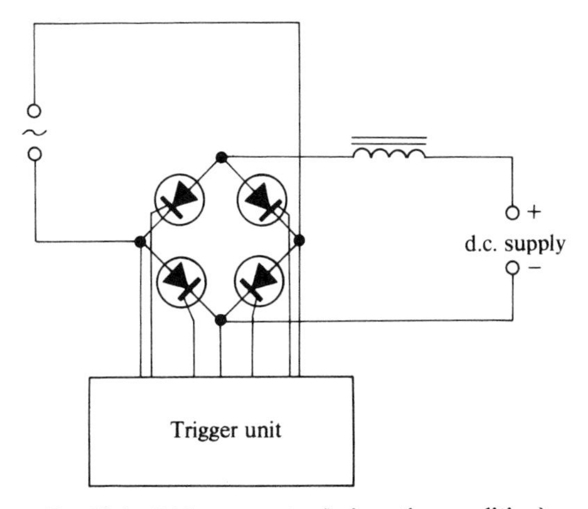

Fig. 13.6 Bridge converter (in inverting condition).

taneously less than, the d.c. source voltage. The current then flows in opposition to the a.c. supply voltage, and the power is from the d.c. side to the a.c. side.

The continuity of the current waveform depends upon several factors, and turn-off may occur naturally, due to the instantaneous current falling to zero, or may be forced by the a.c. supply voltage. This is achieved by pulsing the two thyristors which are instantaneously blocking.

The amount of power inverted depends upon the relative magnitudes of the d.c. and a.c. voltages, and the impedance in circuit, and can be controlled by varying the point in the supply cycle at which the thyristors are pulsed.

13.10 Control of d.c. machines from d.c. supplies

On d.c. systems using thyristors it is necessary to connect a reverse voltage across the thyristor to switch it off. A simple forced commutation system is shown in Fig. 13.7. The circuit relies for its operation upon the fact that TH 1 and TH 2 are not allowed to carry forward current simultaneously. Resistor R has a high resistance and the power consumption in this resistor is insignificant.

To determine the size of commutating capacitor C:

$$C = \frac{t_o I_A}{V_C}$$

where C = commutating capacitance

t_o = thyristor turn-off time

I_A = anode current before commutation

V_C = voltage on C before commutation.

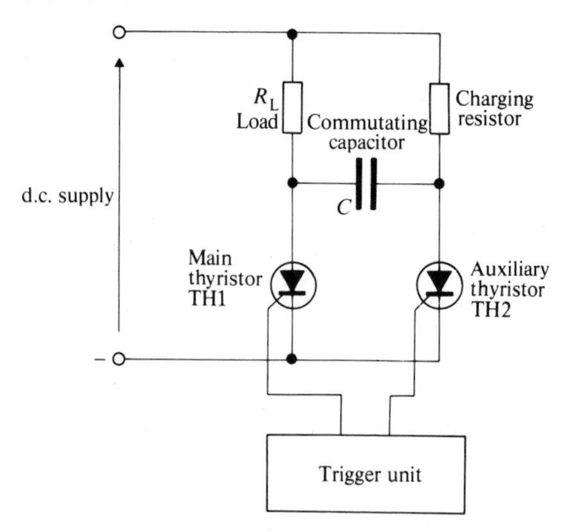

Fig. 13.7 Simple forced commutation.

Thus, for C to be as small as possible, it should be charged to as high a voltage as possible, and the thyristor selected to have as low a value of t_o as possible.

For example, $t_o = 20\ \mu s$; $V_c = 36$ V; $I_A = 200$ A

then

$$C = \frac{20 \times 200}{36} \approx 110\ \mu F$$

When TH 1 is triggered, the supply voltage is switched across the load, and the capacitor C is charged to this voltage via resistor R. If TH 2 is now fired, the capacitor voltage is instantaneously impressed in reverse across TH 1, which switches off – thus switching off the load R_L. When TH 1 is again fired TH 2 becomes switched off, and the sequence is repeated.

13.11 Chopper techniques

One of the most effective means of controlling the speed of a d.c. motor is to vary the average armature voltage. When the supply is already d.c., e.g., battery driven electric vehicle, this can create a problem. This problem could be overcome by using resistors, but this would result in high power losses – creating an inefficient control system. A more efficient solution to the problem is to *chop* the d.c. battery voltage to supply pulses of current to the load as shown in Fig. 13.8.

Many chopper circuits have been devised, and a popular thyristor chopper circuit is shown in Fig. 13.9, in which TH 2 is fired to cause capacitor C to

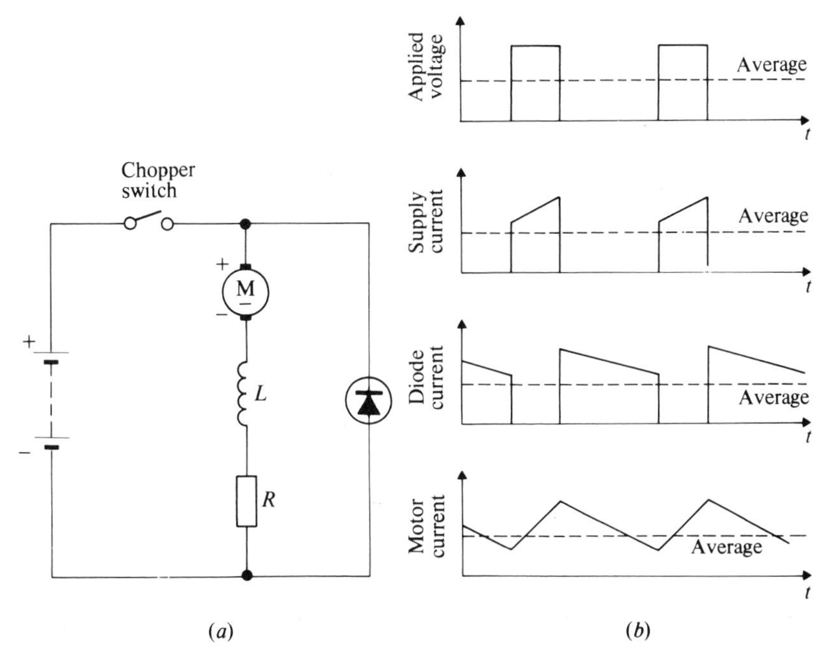

Fig. 13.8 Basic chopper control for motor operation.

charge to the supply voltage with its upper plate positive. TH 2 is now turned off. Thyristors TH 1 and TH 3 are fired to start the load cycle. The motor voltage rises to the supply level, and capacitor *C* discharges through TH 3 and inductor *L*. This is an oscillatory circuit, and *C* charges up with its lower plate positive. TH 2 is again fired, and this connects the capacitor voltage across TH 1, thus turning it off. Capacitor *C* discharges through the load and battery and recharges with its upper plate positive ready for the next sequence.

Chopper circuits may be made by using power transistors to switch the supply current.

Practical Exercise 13b

Simple d.c. motor speed control from a.c. supply

Connect up the circuit shown in Fig. 13.10, using a 13 V a.c. supply to control the speed of a 12 V d.c. series motor.

Observe the waveforms of voltage across the motor armature and the a.c. supply voltage using a dual-trace CRO. Vary the resistance R and note the effect on the motor speed and upon the waveforms.

Connect the flywheel diode across the motor armature and observe the effect on the motor voltage waveform.

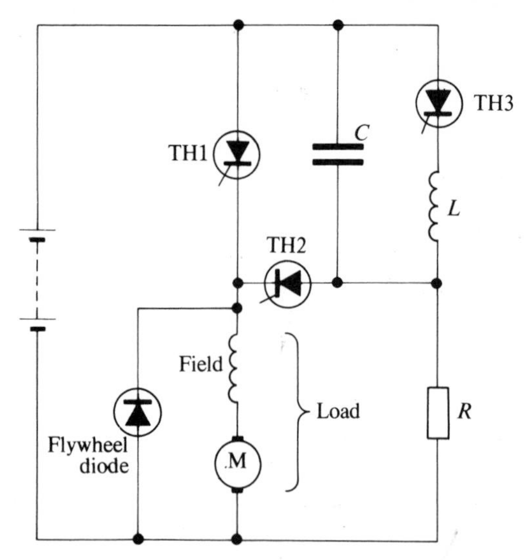

Fig. 13.9 Thyristor d.c. chopper to control speed of d.c. series motor.

Practical Exercise 13c

D.C. motor speed control using a transistor chopper

Connect up the circuit shown in Fig. 13.11. Switch on the 12 V supply and observe the effect on the speed of the motor as the resistor R is adjusted.

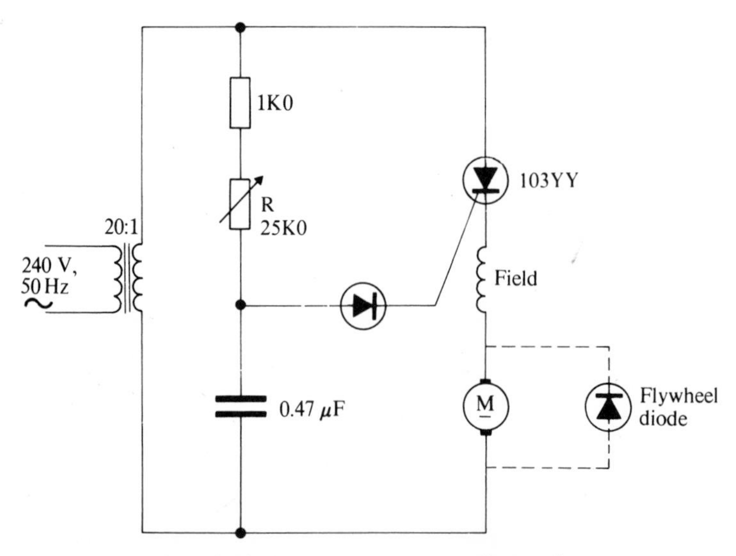

Fig. 13.10 Simple d.c. motor speed control.

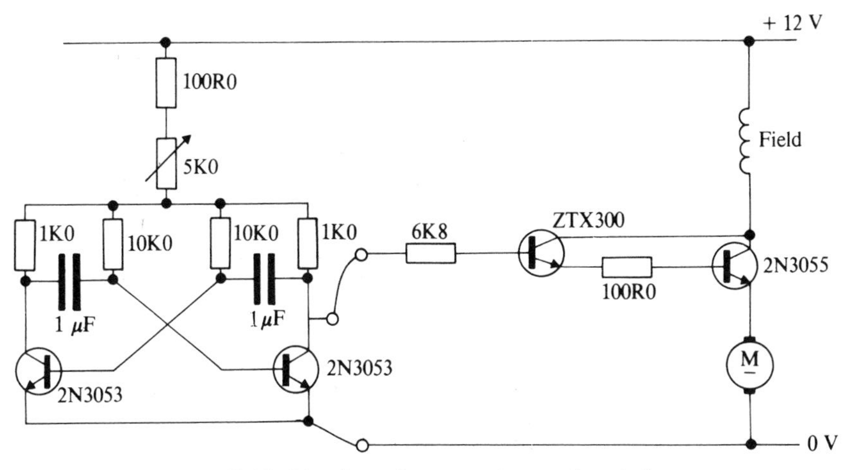

Fig. 13.11 Transistor chopper motor speed control.

In this system, one of the outputs from an astable multivibrator is applied to the servo amplifier input, so that the input current is effectively being chopped.

Practical Exercise 13d

D.C. motor speed control using a thyristor chopper

Connect up the circuit shown in Fig. 13.12. Switch on the 12 V supply and observe the effect on the motor speed as the resistor R is adjusted.

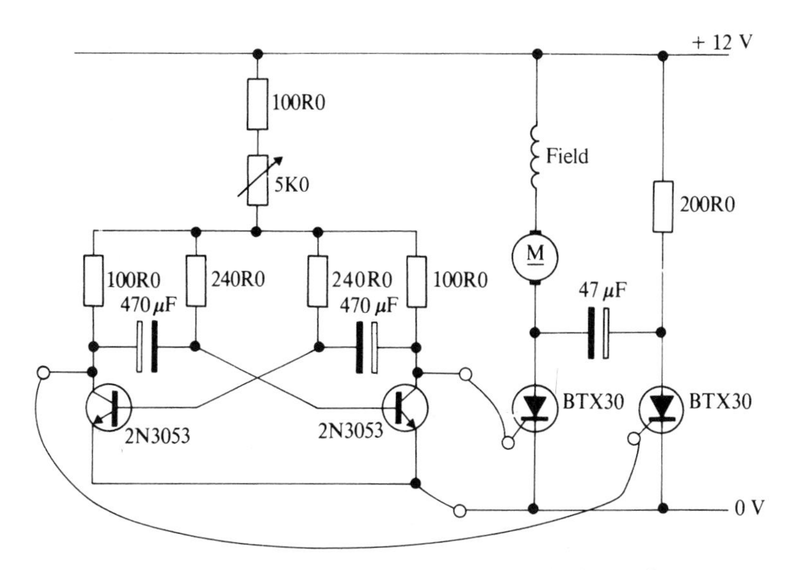

Fig. 13.12 Thyristor chopper motor speed control.

In this system, the two outputs of the astable multivibrator are used as the trigger inputs to two thyristors – thus enabling commutation of the thyristors to be easily accomplished.

13.12 Speed control of a.c. induction motors

The factors which affect the speed of induction motors are: number of poles, supply frequency, supply voltage, load. Speed control may be achieved by varying the supply voltage if the supply frequency and the number of poles are fixed. However, this also affects the torque (torque is proportional to voltage squared), which, in turn affects the amount of load the motor can drive.

On the other hand, if the supply voltage is fixed, speed control can be achieved by varying the frequency, which can be effected with an inverter. A simple d.c. to a.c. inverter, capable of handling several kilowatts at better than 95 per cent efficiency is shown in Fig. 13.13.

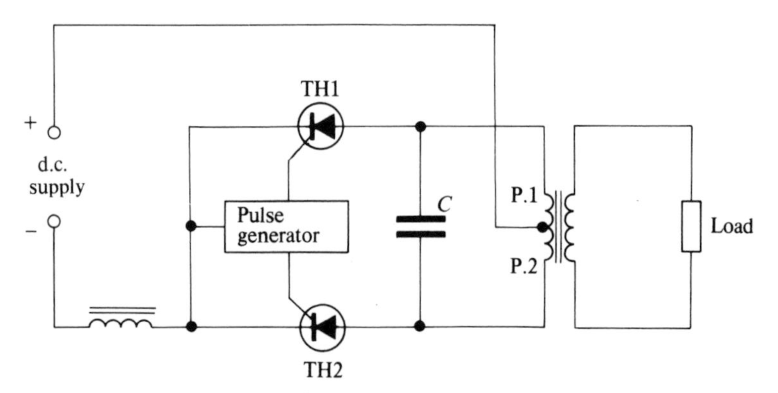

Fig. 13.13 Single phase d.c. to a.c. inverter.

With the circuit initially dead, TH 1 is triggered. Conduction commences, and the voltage across winding P.1 rises. By auto-transformer action, capacitor *C* is charged to twice the voltage P.1. If TH 2 is now triggered, the capacitor voltage is impressed in reverse across TH 1, which switches off. Conduction commutates to TH 2, and the capacitor voltage builds up in the opposite direction. If TH 1 is now triggered, TH 2 is switched off, the output voltage again reverses, and the sequence is repeated indefinitely.

An alternating voltage is thus developed across the load, the frequency of which is determined by the frequency at which the thyristors are alternately triggered.

The primary object of capacitor C is to commutate the current flow between thyristors, but due to the oscillatory nature of the LC circuit, the size of the capacitor also affects the output voltage waveshape. In general, if a good output waveform is required, i.e., approximately sinusoidal, the capacitor must be considerably larger than the minimum required for satisfactory commutation.

14. Electronic Measurements and Fault Diagnosis

14.1 Introduction

Remember that the basic principle of all instrumentation is that *it should not interfere with the system or variable being measured.* For this *ideal* condition to be met, an ammeter is required to have *very low* or *negligible* resistance, and a voltmeter is required to have a *very high* or *infinite* resistance. The instruments available meet these criteria to a greater or lesser extent depending on their sensitivity. However, when an instrument is chosen to measure any variable, the possible effect of its connection must be considered.

The test equipment most commonly used in electronic servicing is as follows:

1. Multimeter.
2. Insulation tester.
3. Oscilloscope.
4. Signal (function) generator.
5. Transistor tester.

14.2 Multimeters

Analogue multimeters (using a moving-coil meter) have been widely used for many years. Unfortunately, while they may be perfectly adequate for their intended uses in electrical servicing, the sensitivity of this type of instrument is not high enough for making accurate measurements on modern electronic equipment, e.g., high-impedance–low-voltage semiconductor circuits. However, in some areas of the electronics industry, e.g., radio and television, the AVO Multimeter Model 8 is still generally used, and the equipment manufacturers specify test voltages which have already taken into account the loading effect of the instrument.

Electronic instruments were developed to overcome the problems caused by the analogue meters. The first in the range were valve voltmeters, which required time to warm-up before use, and they suffered from a problem known as *drift.* Modern electronic multimeters use semiconductor technology which gives the instrument a high input impedance, typically 10 MΩ. These instruments invariably have a digital display using LEDs or LCDs (liquid crystal display) and have, therefore, become known as *digital multimeters* – or generally DMMs.

Very few precautions are necessary when using DMMs due to the advanced technology employed, but you are always advised to check in the manufacturer's handbook before making any measurements. In the case of many of the less expensive DMMs it is necessary to take special care when:

1. measuring currents in excess of 100 mA;
2. making current measurements in circuits with inductive loads;
3. making voltage measurements on high impedance circuits, e.g., MOS and CMOS.

14.3 Insulation tester

It is frequently necessary to test the insulation resistance between components in circuits. In circuits operated from the a.c. mains supply, the *Megger* range of insulation testers are widely used. These instruments employ a hand-wound generator to produce the high voltage, which is applied across the two points in the circuit to be tested. An analogue display, calibrated to read resistance, gives a direct indication of the insulation resistance.

This type of instrument would cause a lot of damage if used on most modern semiconductor circuits and its use must, therefore, be avoided. The only resistance measurements which can be made in such circuits can be achieved with the battery-operated ohmmeter, or the DMM which can measure up to 10 MΩ. Remember that when making resistance measurements it is always necessary to disconnect one end of the component being tested from the circuit, since the shunting effect of other components in the circuit will affect the reading.

14.4 Oscilloscope

The oscilloscope, or CRO, is a very useful instrument in electronics, since it allows us to 'see' the variable which is being measured. This instrument also has a high input impedance, typically better than 1 MΩ, which, when suitably calibrated, can be used to make a wide range of measurements.

Oscilloscopes are normally very tolerant instruments, but check the manufacturer's handbook for precautions to be observed for a particular instrument. In many cases, the following precautions should be noted:

1. The maximum voltage which may be safely applied to the Y input is 400 V. Many instruments have a lower limit. Watch it!
2. Avoid having a high intensity fixed spot or line on the screen for a long period – it could burn the phosphor on the screen,
3. On dual-trace oscilloscopes, each input has an earthed lead which must be connected to a common point in the circuit in which measurements are being made – otherwise you may short-out part of the circuit.
4. On dual-trace oscilloscopes, the time-base triggers on *one input only* e.g., input A, or input 1, so this input must be used as the reference and/or single-trace input, in order that a stationary trace may be obtained.

14.5 Signal generator

The signal (or function) generator is widely used in electronic servicing to feed a waveform of known shape, amplitude, and frequency into a circuit so that it may be traced through the circuit in order to test its performance.

Most of these *function* generators are capable of producing *sine, square, triangular*, and *ramp* waveforms at continuously variable frequencies from fractions of a hertz to several megahertz. Additional features can include a *sweep* facility, which is very useful for frequency response testing and display.

It is always advisable to check with the manufacturer's handbook for any special operating precautions, such as the minimum load resistance, typically 50 Ω, that will not overload the generator.

14.6 Transistor tester

The transistor tester is capable of testing the serviceability of transistors and diodes – which can, of course, be accomplished using an ohmmeter, as described in Chapters 5 and 6. However, this instrument also measures other transistor parameters, in particular h_{FE} (the current gain) which the ohmmeter will not measure.

Many transistor testers have been manufactured, and the handbook should be consulted before using the instrument.

14.7 Techniques of servicing

Fault diagnosis of electronic equipment employs certain standard techniques, each of which we shall briefly consider:

1. ***Functional flow testing.*** This test is widely used in multistage electronic systems or circuits to identify the stage in which the fault has occurred. The faulty stage would then be tested using the specification test. In the functional flow test a signal is applied to the input and the output waveform of each stage is measured and compared with that expected from a serviceable unit.

   In complex systems, such as that shown in Fig 14.1, the *half-split* method is used. Here, the first check is made in the middle of the system, e.g., the output of stage D. If the waveform is correct at this point, then the fault lies in the remaining half of the system. The second half of the system is then checked at its middle, e.g., the output of stages E and F, and so on, until the faulty stage is located. If the expected waveform is not obtained at the output of stage D, then the fault must be in a stage prior to D.
2. ***Specification testing.*** Although this technique may be used to check a complete system, it is much more widely used test the single faulty stage located by the functional flow test.

   In this technique, measurements are made and compared with the stan-

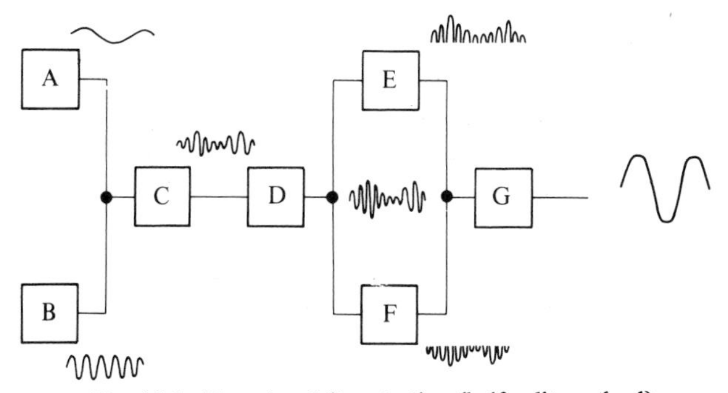

Fig. 14.1 Functional flow testing (half-split method).

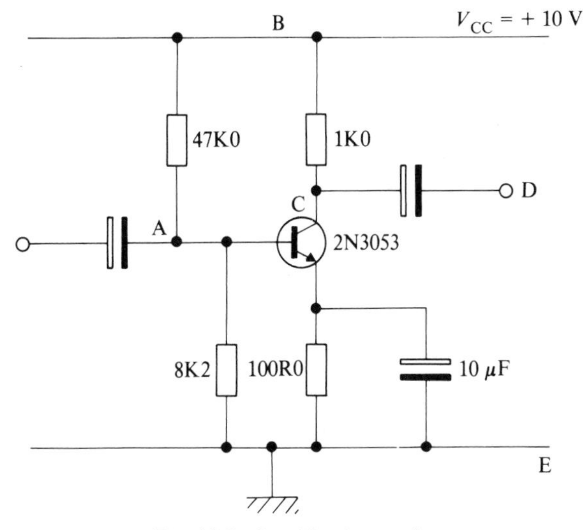

Fig. 14.2 Specification testing.

dard readings for that stage. D.C. voltage measurements are made first as shown in Fig. 14.2, since any incorrect d.c. bias voltage will inevitably upset the a.c. waveform when the signal is applied.

The standard values and measured values of d.c. voltage for the circuit shown in Fig. 14.2. are as follows:

	Standard values (V)	Measured values (V)
A	+1.1	+0.7
B	+10	+10
C	+6.0	+3.1
D	+0.4	+0.0
E	0	0

The faulty component could therefore be the 10 μF emitter by-pass electrolytic capacitor. Replace the capacitor and repeat the measurements. Finally, apply the a.c. signal to the input of the stage and check that the output waveform conforms to the standard.

3. ***Algorithm testing.*** This method of testing is applied to large-scale installations, where a particular fault may repeatedly occur and a collection of statistical results can be obtained. The algorithm is a form of diagnosis which is based on observations of symptom patterns, from which a type of flow chart can be drawn, as shown in Fig. 14.3.

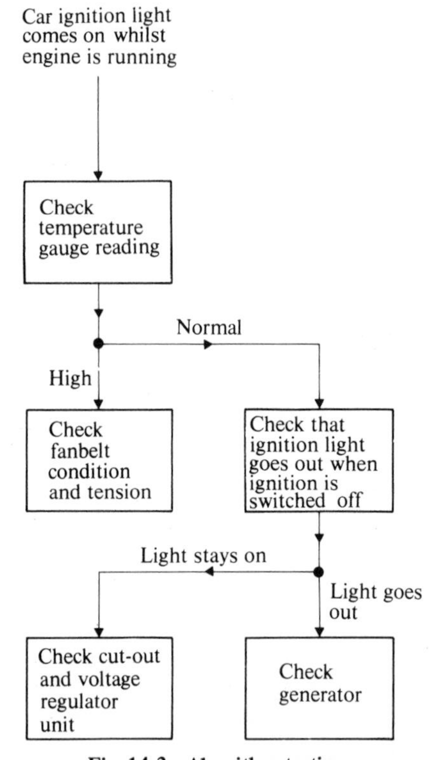

Fig. 14.3 Algorithm testing.

14.8 Summary of measurements in electronic servicing

1. ***D.C. voltage.*** DMM, high accuracy, digital readout. CRO, lower accuracy, more difficult to read.
2. ***A.C. voltage.*** DMM, high accuracy, digital readout, r.m.s. values (assuming a sinusoidal input). CRO, lower accuracy, peak-to-peak value easily read, waveform is visibly displayed.
3. ***A.C. and d.c. current.*** DMM, generally introduces some resistance into the

circuit. It is for this reason that we avoid direct measurement of current, but make a voltage measurement across a known resistance, and then $I = V/R$.

4. ***Frequency.*** CRO, approximate values (often good enough) using the calibrated time base. More accurate measurements can be made using Lissajous figures (see Sec. 14.9). Timer counter, digital readout. This instrument is generally very expensive and can only be justified where very accurate frequency measurement is required – particularly for repetitive measurements.

Practical Exercise 14a

Voltage measurement

Connect up the circuit arrangement shown in Fig. 14.4.

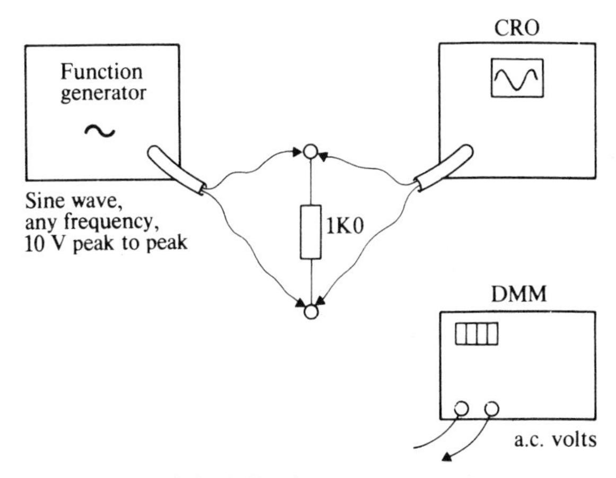

Fig. 14.4 A.C. voltage measurement.

Set the function generator to SINE wave, frequency 10 kHz and output amplitude to max, i.e., 10 V peak-to-peak.

Measure the frequency of the voltage across the 1K0 resistor using the calibrated time base on the CRO.

Note: Do not connect more than one instrument at a time across the 1K0 resistor, because of the loading effect of one instrument on another.

Measure the voltage across the 1K0 resistor with each of the instruments in turn and record the results:

1. CRO peak-to-peak voltage = ________V.
   Therefore,

$$\text{r.m.s. value} = \frac{\text{Peak-to-peak voltage}}{2} \times 0.707 = \_\_\_\_\_\_\_\_ \text{V.}$$

2. DMM voltage = ________V, i.e., r.m.s. value.
3. AVO Multimeter Model 8 voltage = ________V, i.e., r.m.s. value.

Now, set the function generator to SQUARE wave and repeat the measurements.

4. CRO peak-to-peak voltage = ________V.
   Therefore, r.m.s. value = Peak value = ________V.
5. DMM voltage = ________V.
   Therefore, r.m.s. value = DMM reading x 1.11 = ________V, where 1.11 is the form factor (r.m.s./average) for a sine wave and 1.0 is the form factor for a square wave.
6. AVO Multimeter Model 8 voltage = ________V.
   Therefore, r.m.s. value = Reading x 1.11 = ________V.

   Compare the results obtained with the three instruments.

14.9 Special measurements

Several special measurements are possible using the facilities provided on the oscilloscope and function generator:

1. ***Phase.*** The phase difference between two a.c. voltage waveforms can be measured on the CRO. The timebase is switched off and the reference waveform is applied to the X input (often found at the rear of the CRO). The waveform to be measured is applied to the Y input as usual. The pattern which is produced on the screen gives an indication of the phase difference as shown in Fig. 14.5.
2. ***Frequency.*** Frequency measurement, using Lissajous figures, is an accurate method of measuring an unknown frequency by comparing it with an accurate reference frequency. The reference frequency is applied to the X input of the CRO with the time base switched off, and the unknown frequency is applied to the Y input. The patterns (known as Lissajous figures) which are produced on the screen give a measure of the unknown frequency in terms of the reference frequency, as shown in Fig. 14.6.
3. ***Frequency response display.*** This is a technique which uses the sweep facility which is available on many modern function generators. When the 'sweep' is selected, the frequency of the output waveform is continually varied between the minimum and maximum values of the frequency range selected. This enables the response of a circuit or amplifer to be plotted over a range of frequencies. The sweep frequency output waveform is applied to the X input of the CRO, with the time base switched off, and the output voltage waveform of the circuit or amplifier is applied to the Y input. The pattern shown on the screen is a double image of the frequency response curve for the circuit being tested, as shown in Fig. 14.7. The maximum response can be determined by adjusting the frequency dial on the function

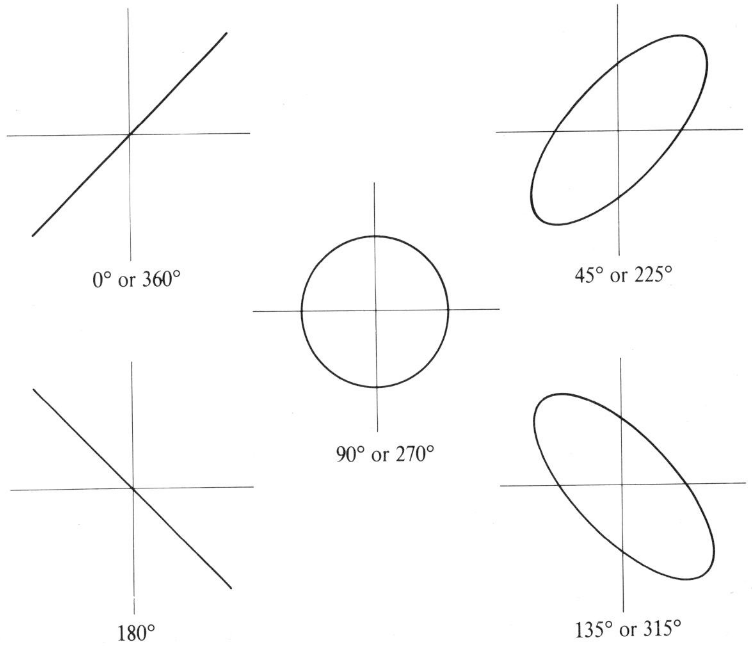

Fig. 14.5 CRO displays for phase differences.

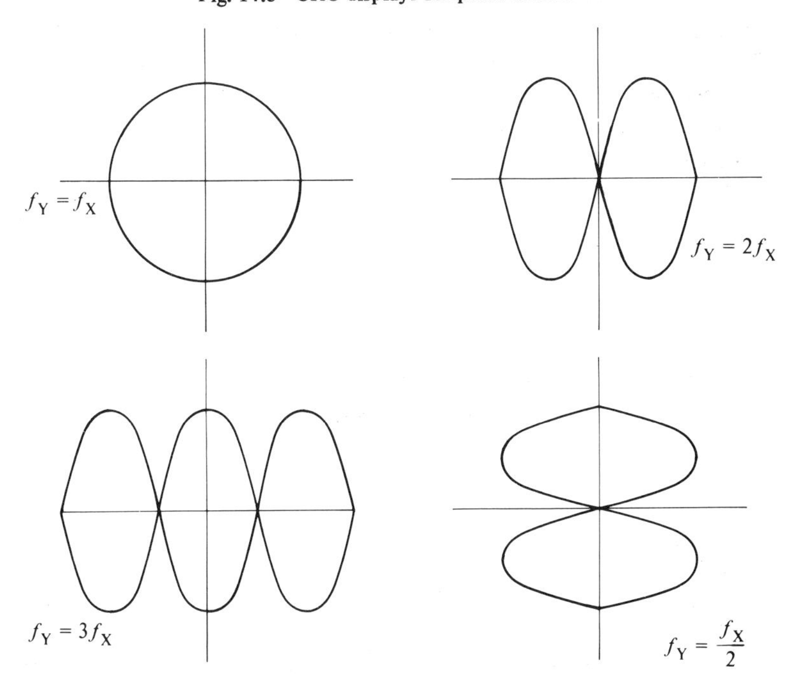

Fig. 14.6 Lissajous figures for frequency measurement.

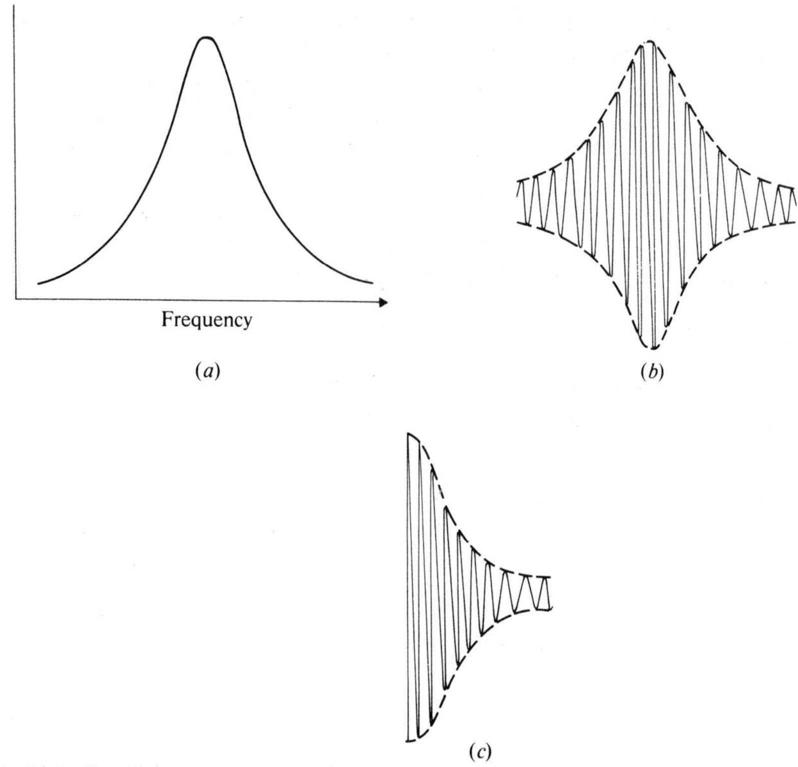

Fig. 14.7 Frequency response waveforms. ***(a)*** **Circuit response;** ***(b)*** **CRO display;** ***(c)*** **CRO display with limited sweep range.**

generator (this limits the range of frequency being swept) until the display is as shown in Fig. 14.7(*c*), when the frequency can be read from the dial.

Practical Exercise 14b

Phase measurement

Connect up the circuit shown in Fig. 14.8.

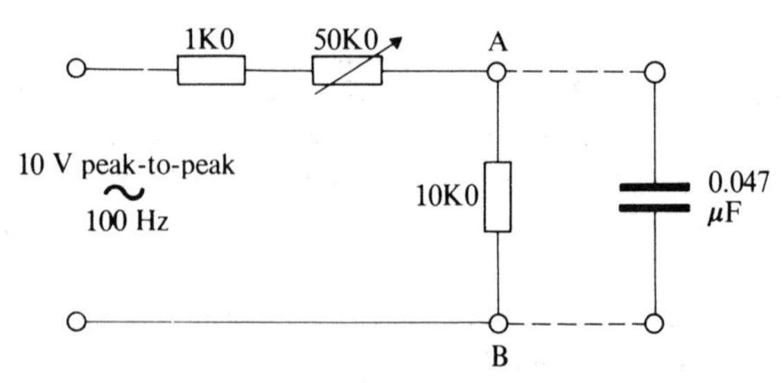

Fig. 14.8 Circuit for phase measurement.

Set the function generator to SINE wave, amplitude 10 V peak-to-peak at 100 Hz and connect the output to the input of the CRO.

Connect the 2.5 V constant amplitude sine-wave output of the function generator to the X input of the CRO. Set the CRO timebase switch to EXT.

Connect the output voltage of the circuit across AB to the Y input of the CRO and observe the display.

Replace the 10K0 resistor by the 0.047 μF capacitor and change the input frequency to 1 kHz. Observe the display on the CRO, and note the effect on the display of varying the 50K0 variable resistor.

Practical Exercise 14c

Frequency measurement by Lissajous figures

Connect up the circuit shown in Fig. 14.9.

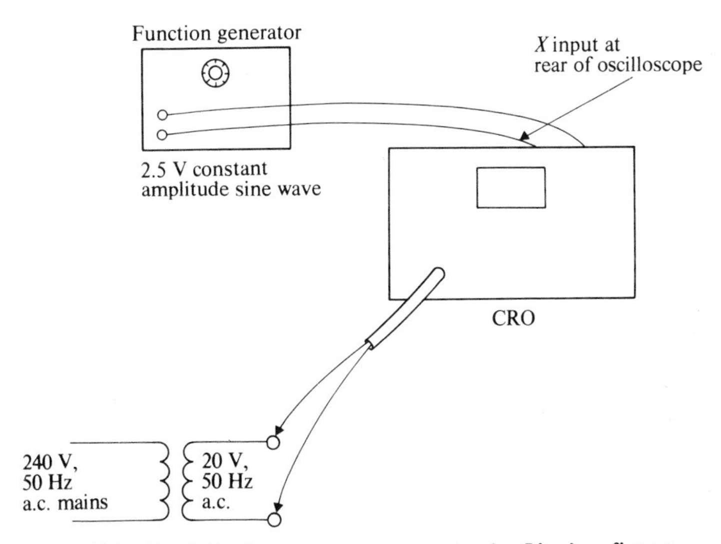

Fig. 14.9 Circuit for frequency measurement, using Lissajous figures.

The reference signal is obtained from the 2.5 V constant amplitude sine wave output from the function generator, and this reference is connected to the X input of the CRO. The CRO timebase switch is set to EXT.

Connect the 50 Hz a.c. mains sine wave output from the transformer to the Y input of the CRO.

Set the function generator to 50 Hz, and adjust the fine control to produce a stationary circle on the screen.

Repeat for frequency settings of 25 Hz, 100 Hz and 150 Hz.

Practical Exercise 14d

Frequency response display

Connect up the circuit shown in Fig. 14.10.

Connect the *sweep rate* ($V \propto f$) output from the function generator to the X input on the CRO, and set the timebase switch to EXT. Apply the normal output from the function generator to the circuit, and connect the output from the circuit (across points A and B) to the Y input on the CRO.

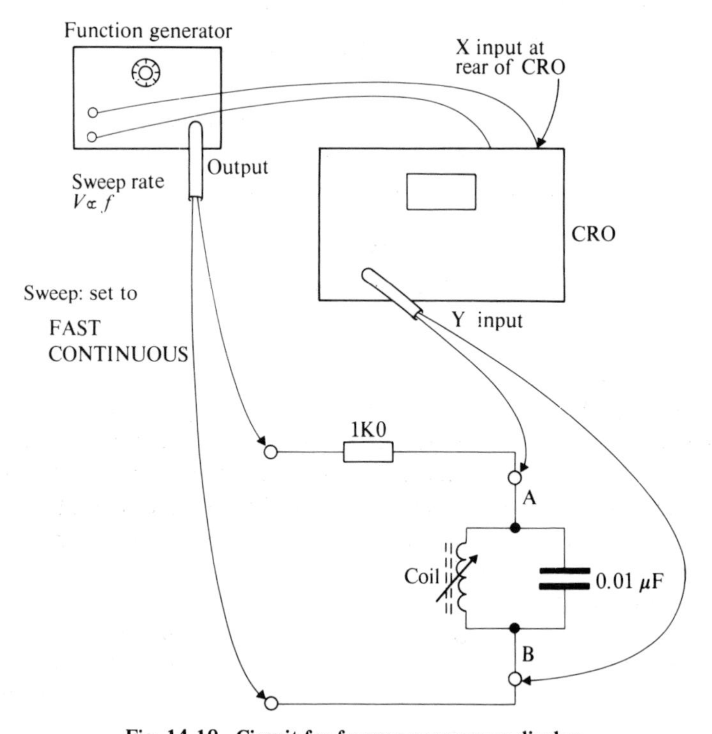

Fig. 14.10 Circuit for frequency response display.

Set the sweep to FAST, CONTINUOUS and adjust the frequency until the response curve is displayed on the screen.

Observe the effect on the display of varying the tuning of the circuit, and the effect of loading the tuned circuit with a 10K0 resistor in parallel.

Vary the frequency dial on the function generator and determine the frequency at which the maximum response occurs.

Practical Exercise 14e

Measurement of input and output impedance of amplifier

Using the amplifier circuit shown in Fig. 14.11, determine the voltage gain.

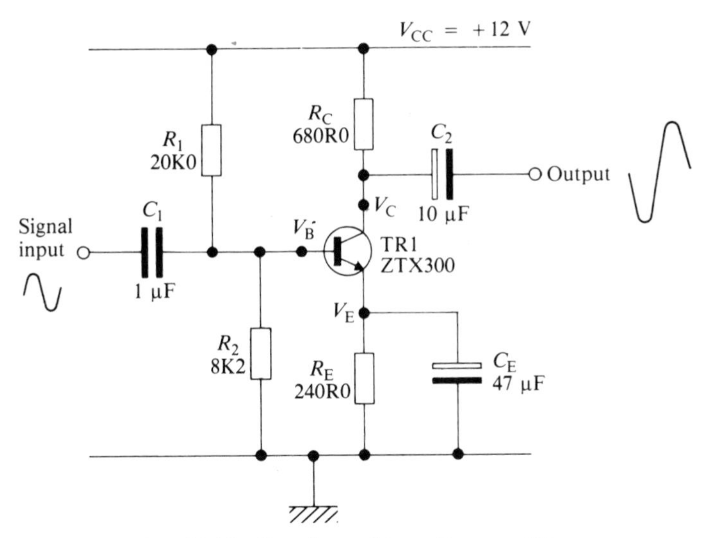

Fig. 14.11 Class A transistor voltage amplifier.

1. ***Input Impedance, Z_{in}.***
   Connect up the circuit arrangement shown in Fig. 14.12(*a*). Adjust the variable resistance R_V to zero ohms. Adjust the amplitude of the input signal to produce the maximum undistorted output. Record this output. Adjust the resistance R_V to decrease the amplifier output by *HALF*.
   Now, R_V and Z_{in} form a potential divider, therefore, assuming linear operation R_V must equal Z_{in}. Measure R_V.
2. ***Output Impedance, Z_{out}.***
   Connect up the circuit arrangement shown in Fig. 14.12(*b*), but with R_V initially open-circuit. Adjust the amplitude of the input signal to give the maximum undistorted output. Record this output. Adjust R_V until the output signal is reduced by *HALF*. R_V and Z_{out} are effectively in parallel, and assuming linear operation, R_V must now equal Z_{out}. Measure R_V.

14.10 Components and common faults

1. ***Resistors.*** Most resistors used in electronic applications are made of carbon film or metal oxide film, so that any damage caused by heat or mechanical stress is likely to cause an increase in resistance or open-circuit.
   Variable resistors may suffer from mechanical wear between the wiper and the track – causing intermittent contact, increased resistance or open-circuit.
   Resistor values may be quickly checked using an ohmmeter (or multimeter, set to ohms).

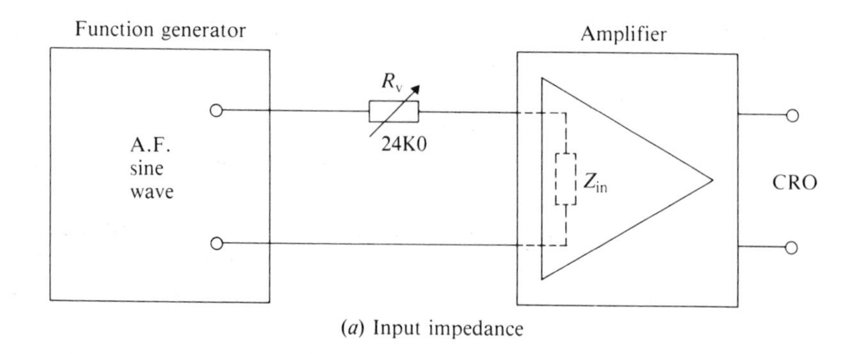

(*a*) Input impedance

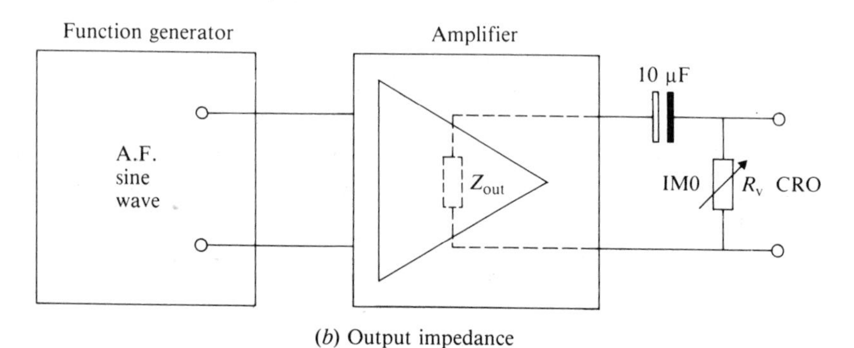

(*b*) Output impedance

Fig. 14.12 Measurement of amplifier impedances.

2. ***Capacitors.*** Electrolytic capacitors are widely used in electronic applications and, since the dielectric is formed by an anodic film, reversal of polarity can cause a breakdown in the dielectric. Reversal to the correct polarity does not always completely recover the film and the capacitor becomes '*leaky*', i.e., a capacitor effectively shunted by a resistance, and the capacitor operates but without complete loss of performance. Leakage current is typically up to about 4 μA. If better than this is required, then it is recommended that solid tantalum capacitors are used. However, extra care must be taken to ensure that the correct polarity is used.

   Low value capacitors (pF up to about 4.7 μF) generally use a solid dielectric. Most problems are caused by mechanical breakage of the leads.

   Capacitors may be quickly checked as described in Practical Exercise 3a.
3. ***Inductors and transformers.*** Faults are normally limited to open and short-circuits of the turns, which may frequently be caused by overheating.
4. ***Semiconductor devices.*** Heat and overvoltage are the main causes of failure.

Simple tests for diodes and transistors are described in Practical Exercise 5a and Practical Exercise 6a respectively.

14.11 Transistor circuit fault diagnosis

It is most important to appreciate that before faults can be diagnosed one must clearly understand the way the circuit works.

Normally, fault diagnosis is achieved by taking voltage readings at predetermined test points and comparing these with the normal readings, then by an analysis of the voltage readings, the component or components at fault can be determined.

In order to understand the operation of a transistor amplifier, consider the operation of the transistor.

Transistors are described by their parameters. One of these, denoted by h_{FE} is the current gain, defined as

$$h_{FE} = \frac{\text{change in collector current}}{\text{change in base current}}$$

For the ZTX 300 n p n silicon planar general-purpose transistor, the typical value of h_{FE} is 100. Substituting this value into the above definition gives:

$$100 = \frac{\text{change in collector current}}{\text{change in base current}}$$

Alternatively,

change in collector current = 100 x change in base current.

This means that if the base current is changed by 1 μA then the collector current changes by 100 x 1 μA = 100 μA.

Therefore:

Golden rule one. A small change in base current causes a large change in collector current.

Golden rule two. An increase in base current gives an increase in collector current – a decrease in base current gives a decrease in collector current. These two rules embody the principle of action of a transistor. That is, a small change in base current causes a large change in collector current. A small current controls a large current.

Practical Exercise 14f

Transistor golden rules, and h_{FE}

Connect up the circuit shown in Fig. 14.13. Vary the base current, and measure base current and collector current for a range of settings of the 1M0 base resistor.

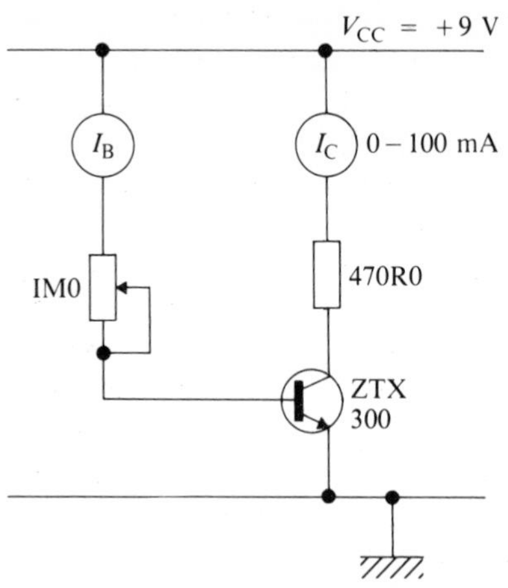

Fig. 14.13 Transistor operating 'golden rules'.

I_B(mA)					
I_C(mA)					

Plot a graph of these results – I_B horizontally, and I_C vertically. From the graph, find the change in collector current corresponding to a change in base current, and determine h_{FE}.

Practical Exercise 14g

Transistor as a switch and d.c. lamp dimmer

Connect up the circuit shown in Fig. 14.14. With the switch S closed, and the

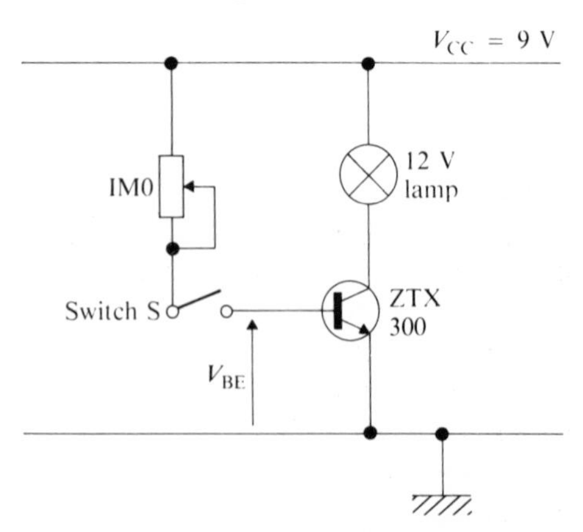

Fig. 14.14 Transistor as a switch and d.c. lamp dimmer.

base resistor adjusted so that the lamp glows, when the switch is opened the lamp will go out. Clearly when the switch is open, $I_B = 0$, so I_C must be equal to 0, hence the lamp will not glow. If the switch is then closed, and there is a base current, this causes a collector current to flow and hence the lamp glows. Adjusting the base resistance changes the base current and hence the collector current. If the base resistance is increased, the base current will decrease, the collector current will decrease and the lamp dims. Adjust the 1M0 resistance so that the transistor is switched HARD ON. Measure the collector–emitter voltage – this should be about 20 mV. Measure the base–emitter voltage (V_{BE}) for the above conditions. It can be observed that when V_{BE} is less than 0.7 V the lamp does not glow, but for V_{BE} greater than 0.7 V the lamp does glow. Thus, by increasing V_{BE}, I_C increases. This means that the collector current I_C is dependent upon I_B *and* V_{BE}. From the observations made in Practical Exercises 14f and 14g it is clear that:

1. a large change in collector current is caused by a small change in base current;
2. as the base current increases so collector current increases;
3. as the base current decreases so the collector current decreases;
4. if the base–emitter voltage increases then the collector current will increase;
5. if V_{BE} decreases then I_C will decrease.

Symbolically, these effects may be represented as follows:

1. $h_{FE} = \dfrac{I_C}{I_B}$
2. If I_B $\uparrow$ then I_C $\uparrow$
3. If I_B $\downarrow$ then I_C $\downarrow$
4. If V_{BE} $\uparrow$ then I_C $\uparrow$
5. If V_{BE} $\downarrow$ then I_C $\downarrow$

Practical Exercise 14h

Transistor thermal effects

1. Connect up the circuit shown in Fig. 14.15.
2. Connect a voltmeter on the 10 V range across the transistor (V_{CE}).
3. Adjust the 1M0 base resistance such that V_{CE} is 2 volts.
4. Observe the voltage V_{CE} with time.
5. Blow on the transistor to cool it.
6. Observe the voltage V_{CE}.
7. Use a match to heat the transistor.
8. Observe the voltage V_{CE}.
9. Explain why the voltage V_{CE} now falls.

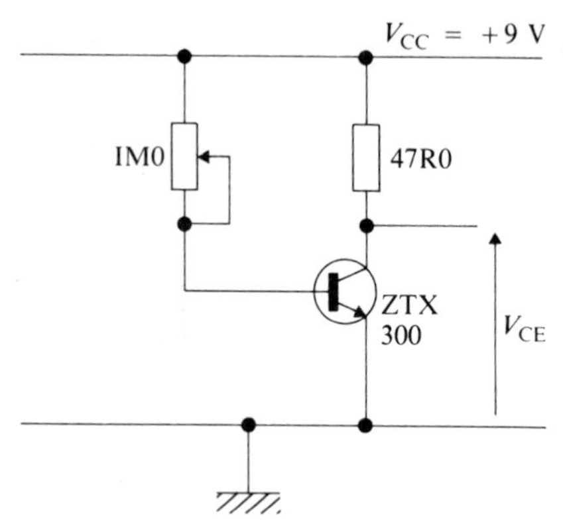

Fig. 14.15 Transistor thermal effects.

10. If your hands are warm put a finger on the transistor.
11. Observe the voltage V_{CE}.

This exercise is likely to pose more questions than it answers. Answers, together with reasons are considered in the next paragraphs.

14.12 Thermal runaway

$$I_E = I_B + I_C$$

But, if $h_{FE} \approx 100$, then $I_B = \dfrac{I_C}{100}$

or, I_B is small compared with I_C; then to a reasonable approximation

$$I_E = I_C$$

Also to a reasonable approximation the power dissipated by the transistor is:

$$P_D = I_C \times V_{CE}$$

Recall that in all electronic devices the power is dissipated in the form of heat.

Next consider the circuit used in Practical Exercise 14h. The two-diode representation of this circuit, with modifications, is shown in Fig. 14.16.

The power dissipated by the transistor is $I_C V_{CE}$.

This power is dissipated as heat. This will cause an increase in the leakage current I_{CBO} of the collector–base diode.

I_{CBO} joins with I_B, thus the effective base current has increased; this causes an increase in collector current (this increase is many times greater than the increase in I_B).

With the increase in collector current there will be an increase in power dis-

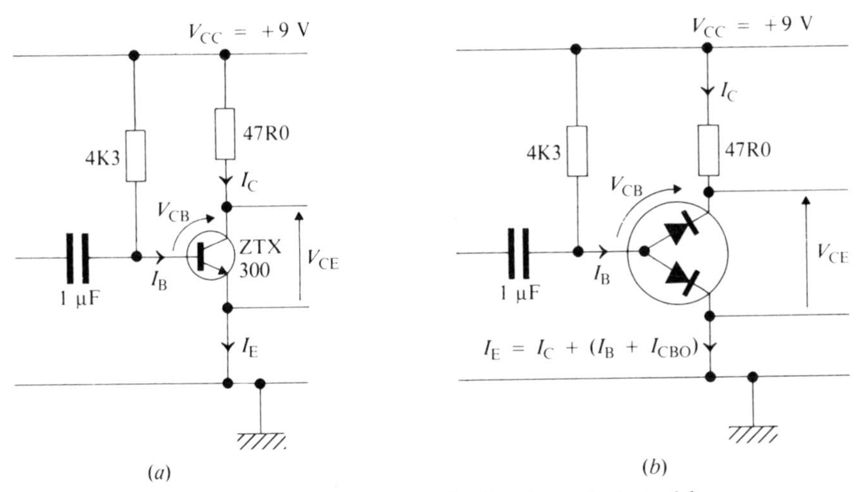

Fig. 14.16 Simple transistor circuit and two-diode model.

sipated by the transistor and hence an increase in temperature. This increase causes an increase in leakage current, and so an increase in effective base current. This will increase the collector current and hence the power dissipated. This causes an increase in temperature and hence an increase in leakage current. The process will continue; thus, with time the collector current and temperature of the transistor will increase. This is called *thermal runaway*, and will eventually cause the transistor to 'burn out' when its junctions reach a temperature of 200 °C.

Since the collector current is increasing, the potential difference across the 47R0 resistor will increase the hence V_{CE} must decrease.

In order to prevent thermal runaway, *feedback* must be introduced to decrease I_B when I_C increases.

As I_C increases, so V_{BC} decreases. Therefore, if the base bias resistor was connected to the collector, then V_{BC} would decrease and hence decrease I_B, thus causing less power to be dissipated and so reduce the temperature of the junction. This method of control is called '*negative feedback*.'

Practical Exercise 14i

Reduction of thermal effects using negative feedback

Connect up the circuit shown in Fig. 14.17. Heat the transistor, e.g., using a match, and compare the rate of decrease of V_{CE} with time, with that obtained in Practical Exercise 14h.

14.13 Minimization of thermal runaway

It has been shown that thermal runaway can have a dramatic effect on an

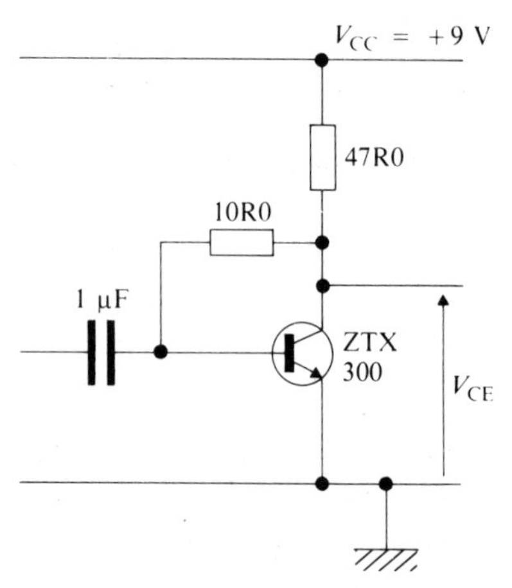

Fig. 14.17 Use of feedback to minimize thermal runaway.

amplifying circuit. Another effect which can cause problems is the large range of h_{FE} for the same type of transistor, e.g., for the ZTX 300 the value of h_{FE} varies between 50 and 300.

Clearly, substituting another transistor having a larger h_{FE} into the circuit will cause the operating output voltage to change, and hence could cause distortion.

These problems can be minimized using the circuit shown in Fig. 14.18.

1. ***Function of R_1 and R_2.*** If 10 times the no-signal base current flows through the potential divider R_1, R_2 to give a voltage V_B relative to the 0 V rail,

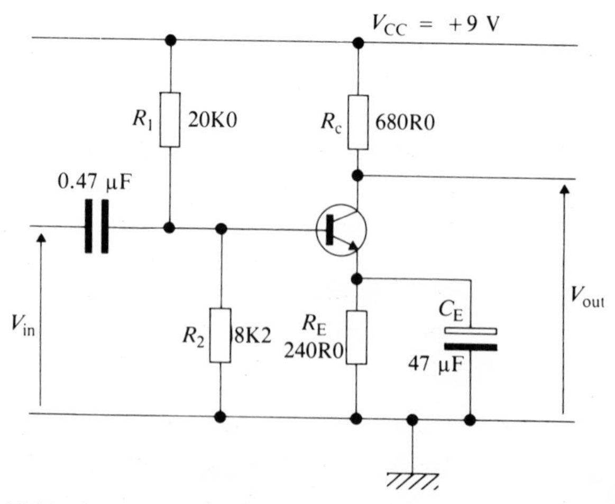

Fig. 14.18 Circuit to minimize thermal runaway and increase stability.

then this voltage will remain almost constant for small changes in I_B. (10 times I_B is a *rule of thumb*, justification can be given by a mathematical analysis of the circuit.)

2. ***Function of R_E and R_C.*** Suppose that due to an increase in temperature or a larger h_{FE} there is an increase in I_C, for the same value of I_B, then $I_E = I_B + I_C$. As there will be an increase in emitter current this will cause an increase in the potential difference across R_E. Since V_B is constant and V_E has increased, then $V_{BE} = V_B - V_E$ must decrease; this, it has been shown, will decrease the collector current. This is a form of negative feedback.

   Should the original transistor need to be replaced by one with a reduced h_{FE}, then the operating voltages will be maintained. If I_C is less, then V_E will decrease and hence V_{BE} increase, thus causing more collector current to flow and so maintaining the operating voltages.

   R_C is the output resistor.
3. ***Function of capacitor C_E.*** Without C_E connected in the circuit, application of a signal at the input would produce very little voltage gain.

   Clearly the NEGATIVE FEEDBACK introduced to prevent other unwanted effects, now destroys the voltage gain.

   As the input signal increases and decreases the base current, so I_E and hence V_E will increase and decrease in phase with the input signal. So as V_B is almost constant and V_E is changing, then V_{BE} must change, its waveform and phase are the same as the signal causing it, so decreasing the collector current and hence the output voltage.

Therefore, a capacitor, whose reactance at the operating frequency, is one-tenth that of the value of the emitter resistance, appears to alternating signals as a low resistance path, whilst the emitter resistor appears as before to the d.c. components.

By this method, *a.c. negative feedback* is reduced and the gain of the circuit is increased. This is the so-called *decoupling* capacitor, C_E.

14.14 Transistor voltage amplifier fault diagnosis

To determine a component fault on a single stage the main diagnosis technique is to use voltage measurements. Voltages are checked against specification sheets, using a 20 kΩ/V voltmeter. It should be recalled that if a semiconductor device becomes open-circuit, then any series connected resistor supplying power to the device will have no voltage drop across it. Similarly, if a semiconductor device becomes short-circuited, then the voltage drop across the series connected supply resistor will increase.

The Electronic Fault Board has been designed to give technicians, having a basic knowledge of semiconductor devices, experience in fault diagnosis on a single-stage transistor amplifier. The Electronic Fault Board is described in Appendix C, together with the p.c.b. layout, component layout, and component list. A local photographer will generally be able to produce a translucent 'positive' from Fig. C.2, which may then be used to produce the p.c.b. Mount the components as shown and solder into place.

The first exercise is to study the circuit under 'normal' operating conditions, and to record a set of 'specification voltages' for these conditions.

Faults are then introduced by setting up combinations of the DIL switches S_a and S_b, which have the effect of opening or shorting-out connections and introducing faulty components.

The reader, when faced with a fault condition, is required to record a set of voltage readings which can then be compared with the 'normal' specification. By applying logical thought processes, the fault should then be confirmed by a few more measurements.

PRACTICAL ELECTRONICS

ELECTRONIC FAULT BOARD—REPORT SHEET

Fault No.:	Switch Selection S_a: S_b:
Voltage readings:	Normal: V_1 = V; V_2 = V; V_3 = V Faulty: V_1 = V; V_2 = V; V_3 = V
DIAGNOSIS AND REASONING:	
SUSPECT COMPONENT AND ACTION TO BE TAKEN:	

Fig. 14.19 Fault report sheet.

To encourage logical diagnosis, the reader is recommended to list his actions and deductions leading to the faulty component. For these purposes, use the form shown in Fig. 14.19. *You are advised to complete the report on one fault before proceeding to take measurements for the next fault.*

Practical Exercise 14j

Electronic fault diagnosis – transistor amplifier

Using the Fault Board described in Appendix C, connected to V_{CC} = + 9 V, set up the switches for 'normal' conditions and measure the voltages using a 20 kΩ/V voltmeter, at each of the test points in the circuit.

Twelve fault conditions may be set-up as shown in the table below:

Fault number	Switch S_a	Switch S_b	
NORMAL	1,5,6,7	1,2,3,6,7	all other switches OFF
1	1,6,7	1,2,3,6,7	
2	1,5,6,7	1,2,3,4,6,7	
3	1,2,5,6	1,2,3,6,7	
4	1,3,5,6,7	1,2,3,6,7,8	
5	1,5,6	1,2,3,6,7	
6	1,5,6,7,8	1,2,3,6,7	
7	1,5,6,7	1,2,3,6	
8	1,5,6,7	1,2,3,7	
9	1,5,6,7	1,3,6,7	
10	1,5,6,7	2,3,6,7	
11	1,5,6,7	1,2,3,5,7	
12	1,3,6,7	1,2,3,6,7	

EXAMPLE 14.1

Consider Practical Exercise 14j (1) as an illustration of the procedure for completing the fault report.

The results are shown in Fig. 14.20.

Continue with the remaining eleven fault conditions, completing a fault report sheet for each fault. It is important to *take your time* – DO NOT RUSH through these exercises, and carefully consider the circuit shown in Fig. 14.18 and Fig. C.3, while attempting to diagnose the fault.

<table>
<tr><td colspan="2">PRACTICAL ELECTRONICS

ELECTRONIC FAULT BOARD—REPORT SHEET</td></tr>
<tr><td>Prac. Ex

Fault No: 14j (1)</td><td>Switch Selection

S_a: 1,6,7 S_b: 1,2,3,6,7</td></tr>
<tr><td>Voltage readings:</td><td>Normal: $V_1 = 2.2$ V; $V_2 = 4.55$ V; $V_3 = 1.6$ V.
Faulty: $V_1 = 0$ V ; $V_2 = 9.0$ V ; $V_3 = 0$ V.</td></tr>
<tr><td colspan="2">DIAGNOSIS AND REASONING:
The potential divider bias network R_1, R_2 is suspect.
With zero volts at the transistor base (V_1), then either R_1 is o/c, OR R_2 is s/c. Assume R_1 is o/c.
Then $I_B = 0$, therefore $I_C = 0$ and $V_{BE} = 0$ V
$V_{RC} = I_C R_C = 0$ V
and $V_{RE} = I_E R_E = 0$ V
Thus: $V_1 = V_{BE} + V_{RE} = 0$ V
$V_2 = V_{CC} - V_{RC} = 9 - 0 = 9$ V
and $V_3 = I_E R_E = 0$ V</td></tr>
<tr><td colspan="2">SUSPECT COMPONENT AND ACTION TO BE TAKEN:
Therefore, the faulty component is diagnosed as R_1 o/c.
Note: Similar effects would be obtained if R_2 was s/c. Therefore, in practice, it would be necessary to pursue further investigation. Firstly, check for ‘dry joints’ at the ends of R_1. If the fault persists, ‘lift’ one end of R_1 and measure its resistance. If R_1 is within the acceptable tolerance band, then ‘lift’ one end of R_2 and measure its resistance. These procedures would confirm that either R_1 is o/c or R_2 is s/c.</td></tr>
</table>

Fig. 14.20 Fault report sheet for Example 14.1.

Appendix A. Units, Multiples, and Sub-multiples

A.1 Fundamental SI units

Quantity	Unit	Unit abbreviation
length	metre	m
mass	kilogramme	kg
time	second	s
electric current	ampere	A
thermodynamic temperature	kelvin	K
luminous intensity	candela	cd
plane angle	radian	rad

A.2 Some derived units

Quantity	Unit	Unit abbreviation
energy	joule	J
force	newton	N
power	watt	W
electric potential	volt	V
resistance	ohm	Ω
inductance	henry	H
capacitance	farad	F
frequency	hertz	Hz
area	square metre	m^2
volume	cubic metre	m^3
linear velocity	metre per second	$m\ s^{-1}$
angular velocity	radian per second	$rad\ s^{-1}$
torque	newton metre	N m
pressure	newton per square metre	$N\ m^{-2}$

A.3 Multiples and sub-multiples

Multiple	Prefix	Meaning
tera	T	Multiply by 1 000 000 000 000 = 10^{12}
giga	G	Multiply by 1 000 000 000 = 10^{9}
mega	M	Multiply by 1 000 000 = 10^{6}
kilo	k	Multiply by 1000 = 10^{3}

Sub-multiple	Prefix	Meaning
milli	m	Multiply by 1/1000 = 10^{-3}
micro	μ	Multiply by 1/1 000 000 = 10^{-6}
nano	n	Multiply by 1/1 000 000 000 = 10^{-9}
pico	p	Multiply by 1/1 000 000 000 000 = 10^{-12}

Appendix B. Greek Alphabet

alpha	A	α
beta	B	β
gamma	Γ	γ
delta	Δ	δ
epsilon	E	ϵ
zeta	Z	ζ
eta	H	η
theta	Θ	θ
iota	I	ι
kappa	K	κ
lambda	Λ	λ
mu	M	μ
nu	N	ν
xi	Ξ	ξ
omicron	O	o
pi	Π	π
rho	P	ρ
sigma	Σ	σ
tau	T	τ
upsilon	Υ	υ
phi	Φ	ϕ
chi	X	χ
psi	Ψ	ψ
omega	Ω	ω

Appendix C. Manufacture and Assembly of Electronic Fault Diagnosis p.c.b.

C.1 Printed circuit board (p.c.b.)

Printed circuits generally provide an economical method for mounting and interconnecting ICs and discrete components. Standard prototype boards are widely available which are developments of the stripboard, in which copper track is bonded to SRBP board or epoxy glass material conforming to a range of requirements, e.g., a general layout suitable for a 'mix' of discrete components and DIL ICs and may also be equipped with 'pads' for edge connection. Other standard circuit boards have tracks and pads bonded on the board to layouts which conform to widely used microprocessor bus systems.

P.c.b.s are produced using several different techniques, the most popular for small-scale production being the *direct method* and the *photographic method*. The former uses a copper-clad epoxy glass board on which the *artwork* is applied directly, either by using a pen filled with etch-resist ink or by using the wide range of available etch-resist p.c.b. transfers (including pads, tracks, connectors, right-angled D connectors, DIL pads, transistor pads, links, etc.). The board, with its artwork of the p.c.b. layout required attached, is then etched in a solution of ferric chloride. After etching, the resist ink and/or transfers are removed, and the board may be drilled ready for component mounting. With the photographic method, special photo-resist copper-clad board is used. The artwork is prepared by applying p.c.b. resist transfers and/or resist ink onto a sheet of transluscent polyester film—alignment of the layout may be aided by using a transparent 0.1 inch layout grid. This artwork is then placed against the positive photo-resist surface of the board and exposed to UV light for several minutes, then removed for developing in a solution of sodium hydroxide. Finally, the board is etched in a solution of ferric chloride. The artwork generally used for this small-scale production is on a 1 : 1 scale. As the degree of complexity of p.c.b. layout increases, it is useful to use larger-scale transfers such as 2 : 1 or even 4 : 1 to prepare the artwork. The enlarged artwork is then *photographically reduced to standard size* for transferring onto the board. This technique achieves a greater accuracy and more easily defined track spacing than is possible using a 1 : 1 scale.

Designers of printed boards for ICs should always aim at obtaining the best possible packing density. Poorly designed, poorly laid out boards cost just as much to produce as well-designed boards, but well-designed boards will minimize interconnection lengths and will generally improve the board's design function. Packing density will be minimized in those applications where the power dissipation is relatively high in order to allow natural air circulation to dissipate the heat generated by the ICs. The designer must also consider such factors as ease of fault finding, since the time taken to locate and replace a defective IC may well affect the value and the number of spares to be held, and also the way in which the system is broken down into its sub-assemblies.

Multi-layer p.c.b.s invariably use an epoxy resin glass fibre as the base material, and are made up from a stack of single-sided etched sheets, laminated by heat and pressure to produce a single board. Experience has shown that the best packing densities and the simplest board layouts are achieved using $X-Y$ co-ordinate wiring, where all the tracks on one side of the board generally run in one direction, and the tracks on the other side of the board run at right angles to them.

Various methods are used for the interconnection of the layers of multi-layer boards, one of the most popular being the *plated-through-hole* technique, where a hole drilled through the interconnection pads is subsequently copper-plated. The multi-layer p.c.b. production is relatively expensive and tends to be restricted to specialist highly complex circuits and systems, e.g., airborne computer systems. However, the application of *double-sided* p.c.b.s has increased considerably during the last decade, coincident with the development and application of microprocessor and microcomputer systems, so that the problems associated with the complexity of interconnections are largely overcome by this technique.

C.2 Mounting and connections

The most common method of mounting components into circuit boards is the *insertion* method, where the components have their leads passed through plated or unplated holes in the p.c.b., and then soldered, as shown in Fig. C.1(*a*). Components having round axial leads, round radial leads and flat leads (i.e., most DIL packages) can be easily mounted by this method.

The reliability of an electronic system depends largely on the number of electrical connections involved, the most reliable system being one with the least number of connections. Permanent methods for connection are inherently more reliable than temporary 'disconnect' methods – which invariably introduce another joint in the interface area. Therefore, the designer must attempt to strike the correct balance between reliability, ease of production, and servicing.

The most widely used method of making permanent connections is by *soldering*, which is the technique of joining metal-to-metal by means of a solder

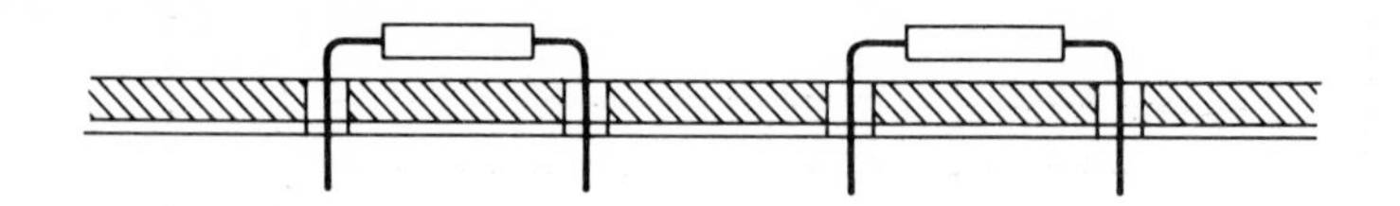

(*a*) Insertion mounting

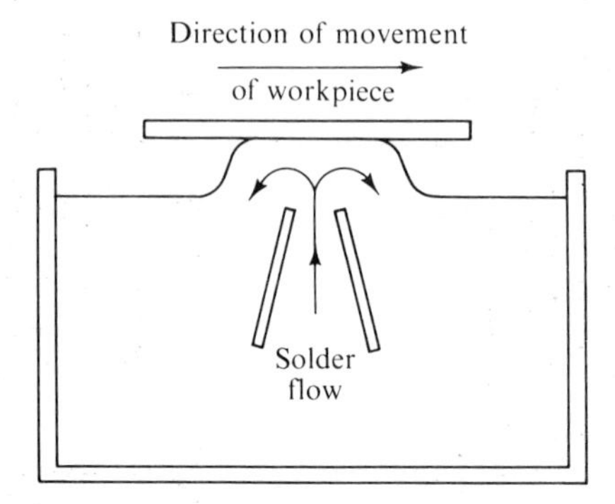

(*b*) Flow soldering

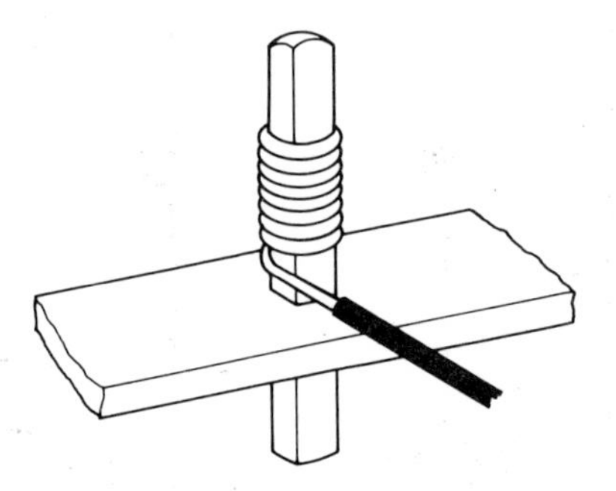

(*c*) Wire wrap connection

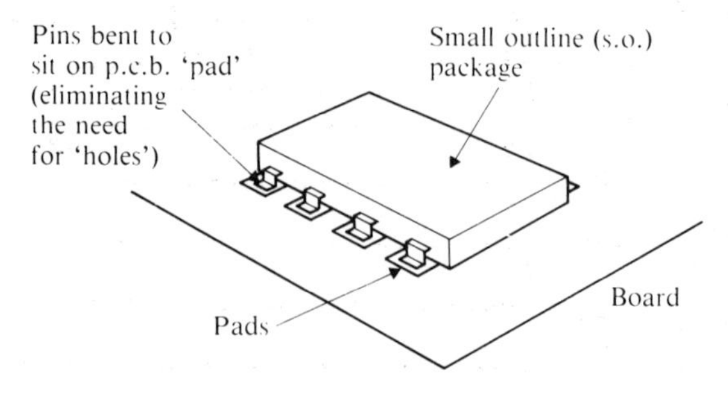

(*d*) Surface mounting

Fig. C.1 Mounting and connections.

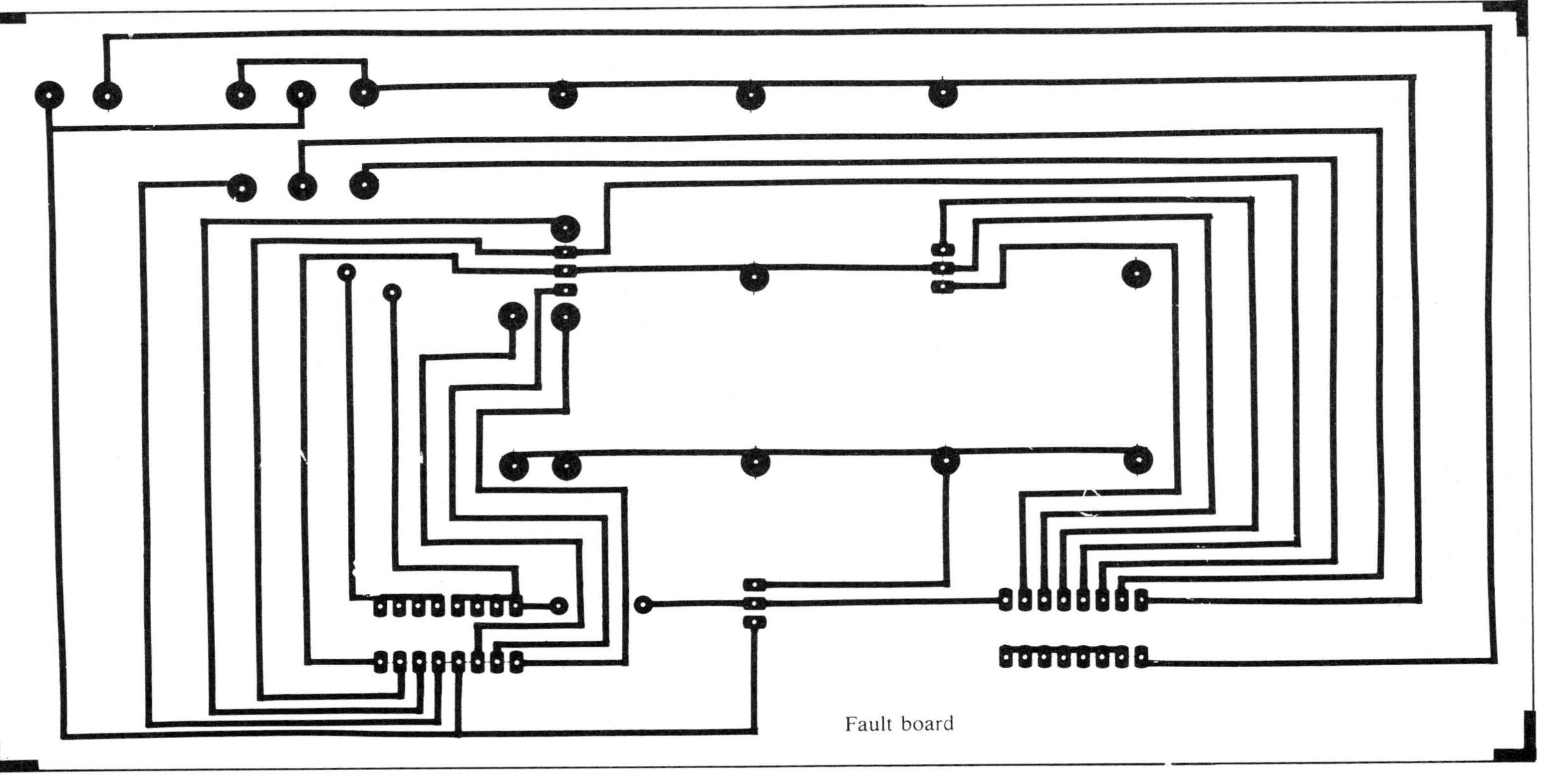

Fig. C.2 P.C.B. artwork for electronic fault board.

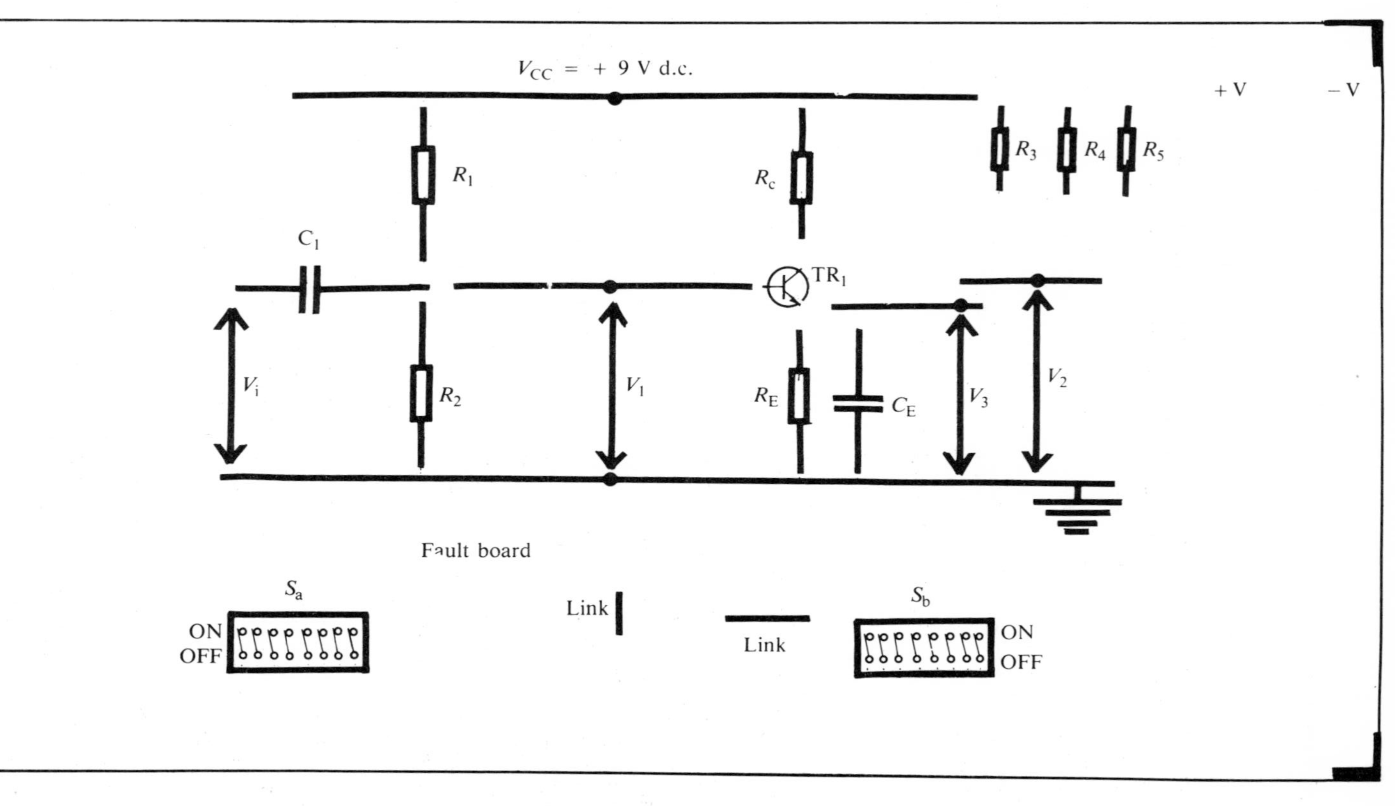

Fig. C.3 Component layout for Electronic Fault Board p.c.b.

alloy employing heat and flux. The flux is used to aid heat conduction from the heat source to the workpiece, and to remove surface oxides at the instant when the molten solder 'wets' (goes into solution) the surfaces to be joined.

Electrical soldering irons are generally used for small-scale production, servicing, repair, and prototype development. Large-scale production tends to favour *flow*, or *wave soldering*, in which boards are passed over a wave of continually circulating molten solder, generated by a special nozzle, as shown in Fig. C.1(*b*). Ideally, the wave should have a flat crest and turbulent sides so that excess solder is removed.

A popular method of interconnection, for hard-wiring and prototype circuit boards, is the wire-wrap connection, since it provides a good quality electrical *and* mechanical connection. Single-strand wire is wrapped around sharp-edged posts under tension, as shown in Fig. C.1(*c*). A minimum of five turns of wire is usually recommended, and can be accomplished using manual or fully automatic electric or pneumatic tools or machines.

Improvements in semiconductor technology and manufacturing methods has enabled the production of much smaller 'chip' size for more complex circuits. This development has reached a stage where the size of package and speed of operation must be questioned. The *small outline (s.o)* was developed in the late 1960s and the principle has been in use for *hybrid* semiconductor packages for many years. Today, increasing numbers of the s.o. IC packages are becoming available with pins which are bent so that the packages are *surface mounted*, as shown in Fig. C.1(*d*). The s.o. package offers about 70 per cent area reduction, eliminates the need for holes in the p.c.b. (and therefore, the problems and high cost of through-hole plating). The lead spacing on IC packages is reduced to 0.03 to 0.05 in compared with the standard 0.1 in of conventional DIL packages.

Surface mounting assembly may use the now-recognized process with solder paste and vapour phase soldering, or the method of glueing the device to the p.c.b. and then wave soldering.

C.3 Electronic Fault Diagnosis – Fault Board p.c.b.

The *artwork* for the Fault Board p.c.b. is shown in Fig. C.2 with the component layout shown in Fig. C.3. You will probably find that your local photographer will produce a *positive* on translucent sheet that you will be able to use to manufacture your p.c.b. Professional p.c.b. manufacturers will generally produce artwork for the component layout (Fig. C.3) and produce a *silk screen* print on the component side of the p.c.b. It is not considered necessary for you to attempt silk screen printing. Reference may be made to Fig. C.3 for identification of components and test points. Resistors R_3, R_4, and R_5 are components used to switch-in faults.

C.4 Fault Board component list

TR 1 = ZTX 300
R_1 = 20K0 ± 5 per cent 0.5 W
R_2 = 8K2 ± 5 per cent 0.5 W
R_C = 680R0 ± 5 per cent 0.5 W
R_E = 240R0 ± 5 per cent 0.5 W
C_1 = 0.47 μF
C_E = 47 μF
R_3 = 4K7 ± 5 per cent 0.5 W
R_4 = 2K2 ± 5 per cent 0.5 W
R_5 = 10R0 ± 5 per cent 0.5 W
S_a, S_b = 8-way S.P. DIL switches.

Appendix D. Typical Results for Practical Exercises

D.1 Chapter 1. Instrumentation

Practical Exercise 1a

The resistance scale is non-linear, being zero at f.s.d., and high at zero deflection:

R (Ω)	0	470	1K	2K2	4K7	8K2	10K	15K	20K	47K	100K
A (div)	100	97	94	88	78	66	62	51	44	26	14

R (Ω)	150K	330K	470K	560K	1M
A (div)	10	5	4	3	2

An unknown resistor gave a deflection of 83 μA divisions, and from the above table we can only say that the resistance is between 2K2 and 4K7.

These results are best represented graphically – using a logarithmic scale for resistance (enabling resistance to be read accurately on the expanded axis), as shown in Fig. D.1, from which the unknown resistance can be seen to be 3K3.

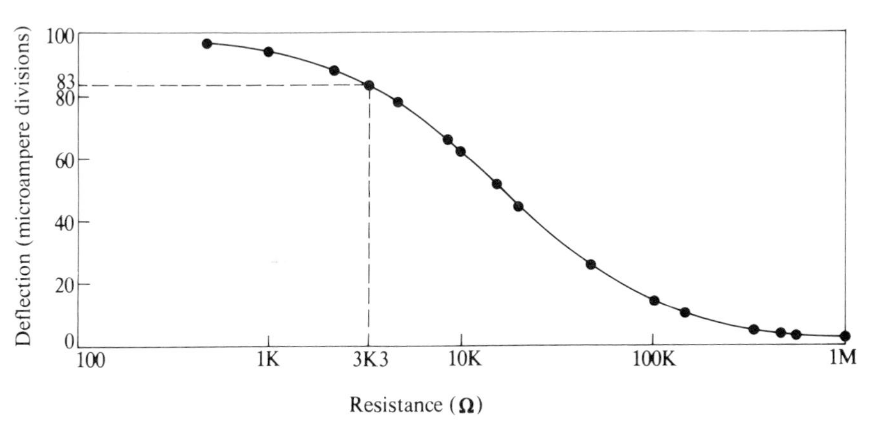

Fig. D.1 Ohmmeter calibration curve.

Practical Exercise 1b

1. Theoretical value of $V_{BC} = 5.0$ V.
2. Resistors both 1K0, $V_{BC} = 5.0$ V
3. Resistors both 100K0, $V_{BC} = 4.0$ V.
4. Resistors both 1M0, $V_{BC} = 1.9$ V

Practical Exercise 1c

	Resistance		
	1K0	100K0	1M0
Theoretical value of V_{BC}	5.0 V	5.0 V	5.0 V
Analogue voltmeter reading V_{BC}	5.0 V	4.0 V	1.9 V
Digital voltmeter reading V_{BC}	5.0 V	5.0 V	4.8 V

From these results, we can conclude that the digital voltmeter is the best, since the readings are closest to the theoretical values – this being due to the high impedance of the DMM not loading the impedance across which the voltage is being measured.

Practical Exercise 1d, part 8

1. When both resistors are 1K0, CRO $V_{BC} = 5.0$ V
2. When both resistors are 100K0, CRO $V_{BC} = 4.75$ V.
3. When both resistors are 1M0, CRO $V_{BC} = 3.3$ V.

D.2 Chapter 2. Resistors

Practical Exercise 2a

1. $V_1 = 2$ V; $V_2 = 3$ V; $V_3 = 4$ V
2. Sum of voltage drops *equals* the supply voltage.
3. $R_T = 45$ kΩ
4. $I = 9/45 = 0.2$ mA.

Practical Exercise 2b

1. $V = 9$ V
2. $I_1 = 0.9$ mA; $I_2 = 0.6$ mA; $I_3 = 0.45$ mA
3. $I_T = 1.95$ mA
4. $R_T = 9/1.95 = 4.62$ kΩ
5. $$R_p = \frac{15 \times 10}{15 + 10} = \frac{150}{25} = 6 \text{ k}\Omega$$

   Therefore

   $$R_T = \frac{6 \times 20}{6 + 20} = \frac{120}{26} = 4.62 \text{ k}\Omega$$

Practical Exercise 2c

1. $V_P = 2.1$ V; $V_1 = 6.9$ V
2. $R_P = \dfrac{15 \times 10}{15 + 10} = \dfrac{150}{25} = 6$ kΩ
3. $R_T = 20 + 6 = 26$ kΩ
4. $I_T = 9/26 = 0.346$ mA.

Practical Exercise 2d

1. When light was directed on the photoresistor in the circuit shown in Fig. 2.6(*a*), the voltage V_{BC} decreased from 9.8 V to 4.6 V.
2. When light was directed on the photoresistor in the circuit shown in Fig. 2.6(*b*), the voltage V_{BC} increased from 0.2 V to 5.4 V.
3. The frequency of light variations from the fluorescent tube was found to be 100 Hz.

Practical Exercise 2e

1. The 'cold' resistance of the 12 V 1W lamp = 25 Ω
2. Instantaneous maximum current taken by lamp = 46 mA
3. The initial values were 15 mA, 1.0 V, 15.0 V
4. After 1 min, the values were 45 mA, 12.0 V, and 4.0 V.
5. The 'hot' and 'cold' resistance of the lamp are, therefore, calculated as follows:

   Hot resistance = 267 Ω
   Cold resistance = 25 Ω
   'Hot' resistance of thermistor = 1 kΩ
   'Cold' resistance of thermistor = 89 Ω

D.3 Chapter 3. Capacitors

Practical Exercise 3a

The magnitude of the initial 'kick' of current during capacitor charging increased with the size of capacitor. The time taken to discharge through the voltmeter increased with the size of capacitor.

D.4 Chapter 4. Inductors and Transformers

Practical Exercise 4a

As the 1.5 V d.c. supply is switched off, the Neon glows momentarily.

D.5 Chapter 5. Rectification, power supplies, and smoothing

Practical Exercise 5a

When the lamp lit, the voltage across the lamp was 11.3 V and across the diode was 0.7 V. The brilliance of the lamp was slightly reduced owing to the forward voltage drop across the diode.

When the lamp did not light, the voltage across the lamp was 0 V, and across the diode was 12.0 V.

Using the analogue type of ohmmeter, a *low* resistance reading of 5 kΩ was measured when the anode was positive and the cathode negative. When the diode was reversed, the resistance reading increased to greater than 1 MΩ.

Using the DMM, the readings were 247 kΩ and greater than 1 MΩ, respectively.

Practical Exercise 5b

1. With C_1 excluded, the peak value of voltage measured was 34 V and the d.c. voltage was 21 V.
   With $C_1 = 10\ \mu F$, the d.c. level = 26 V with an a.c. ripple = 14 V.
   When $C_1 = 100\ \mu F$. $V_{BC} = 30$ V with the a.c. ripple = 2 V.
2. The peak value of the a.c. supply applied to the circuit was 37 V. The steady value of d.c. voltage = 24 V with a minimum a.c. ripple voltage of 10 mV peak-to-peak when $C_1 = 100\ \mu F$. $C_2 = 47\ \mu F$, and $L = 5$ H.

Practical Exercise 5c

Some distortion in output was noted – particularly apparent when music was played.

Practical Exercise 5d

R	0	10K	8K2	4K7	2K2	1K0	680R0	470R0
V	16.5	16.2	16.1	16.1	16	15.8	15.6	15.2
I(mA)	0	1.5	1.7	3	7	16	24	34

R	240R0	180R0	100R0	10R0
V	14.5	13.8	12.5	4
I(mA)	64	82	130	450

The regulation curve is shown in Fig. D.2, from which it can be seen to be a straight line (gradually falling) voltage with increased current.

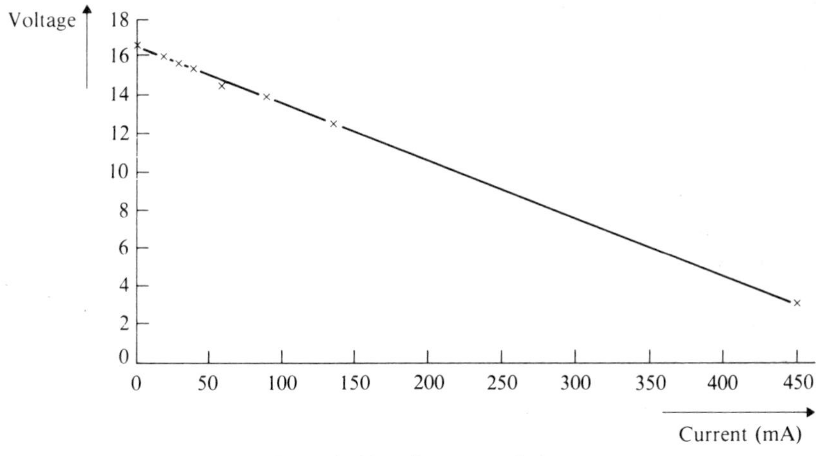

Fig. D.2 Transformer regulation.

Practical Exercise 5e

R	0	10K	8K2	4K7	2K2	1K0	680R0	470R0
V	23	21.8	21.6	21	19.8	18.3	16.8	15.7
I(mA)	0	2.2	2.65	4.5	9	18.2	25.5	33.5

R	240R0	180R0	100R0	10R0
V	12.5	11	8.3	2
I(mA)	54	64	86	200

The graph of current/voltage is shown in Fig. D.3 from which it can be seen that very little improvement has been achieved with regard to regulation.

Practical Exercise 5f

1. $V_{out} = 40$ V
2. $V_{C1} = 20$ V; $V_{C2} = 20$ V
3. $V_{C1} + V_{C2} = V_{out}$
4. $V_{out} = 21$ V; $V_{C1} = 10.5$ V; $V_{C2} = 10.5$ V
5. Increased load dramatically reduces voltage output.
6. With 10K load resistor, $V_{out} = 36$ V
   With 1K load resistor, $V_{out} = 16$ V

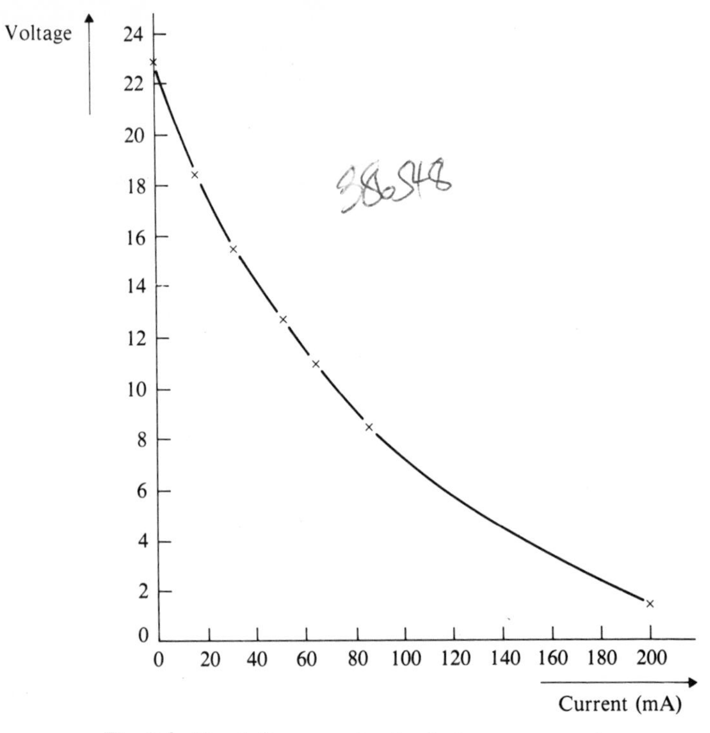

Fig. D.3 Regulation curve for simple d.c. power supply.

Practical Exercise 5g

Forward characteristic:

V (V)	0.2	0.4	0.6	0.7	0.8	0.8	0.8
I (mA)	0.25	0.75	1.5	2.0	5	10	20

Reverse characteristic:

V (V)	10.1	10.2	10.3	10.4	10.5	10.5	10.6	10.7	10.7	10.8
I (mA)	0	1.5	3	5	7	9	10	14	16	17

These results are plotted graphically in Fig. D.4, from which it can be seen that the forward characteristic is similar to a normal silicon diode, and the reverse characteristic has quite a sharp reverse breakdown.

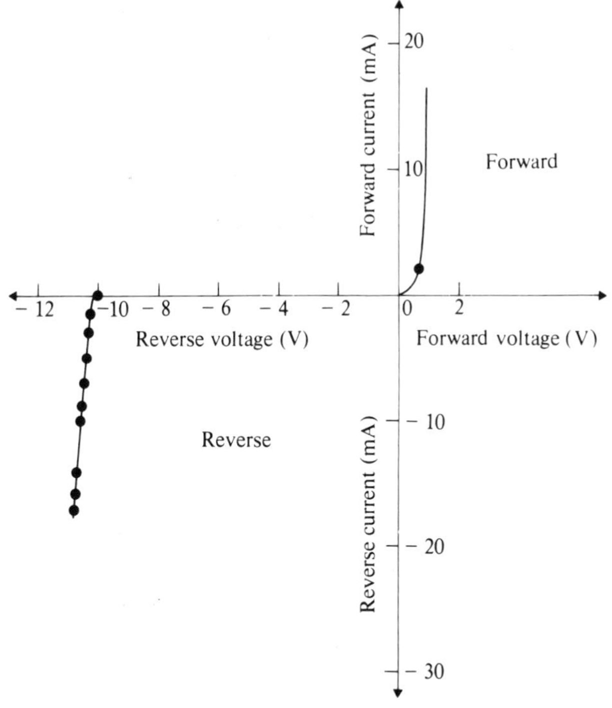

Fig. D.4 Zener diode characteristic.

Practical Exercise 5h

Unstabilized voltage (V)	25	22	20	18	16
Load voltage (V)	10.4	10.3	10.0	4.7	0.6
Zener current (mA)	6	2	0	0	0
Load current (mA)	5	10	15	20	22

From these results it can be seen that the stabilized voltage remains reasonably constant as the load current increases, until a limit is reached when the Zener diode ceases to conduct in the reverse direction.

Practical Exercise 5i

Unstabilized voltage (V)	17	18	20	22	24	26
Load voltage (V)	10.2	10.3	10.3	10.4	10.5	10.6
Zener current (mA)	1	3	5	7	8	10

It can be seen that the stabilized voltage remains reasonably constant over a range of supply variations.

Practical Exercise 5j

Unstabilized voltage, V_1	11	11	11	11	11	11
Stabilized voltage, V_2	3.75	3.75	3.75	3.7	3.5	3.5
Load Current	0.7	2.0	5.0	11.5	20.0	22.0

The stabilized voltage remains reasonably constant over a wide range of load variation.

Practical Exercise 5k

Unstabilized voltage, V_1	11.5	11.3	11.1	10.7	10.2	9.7	9.0
Stabilized voltage, V_2	4.2	4.2	4.2	4.2	4.2	4.2	4.1
Load Current, I (mA)	0.8	2.2	4.6	4.6	6.0	16.4	36.0

Stabilized voltage remains reasonably constant with wide load variation.

Practical Exercise 5l

Minimum load. V_{stab} = 9.2 V → 20.5 V
I = 1.6 mA → 3.6 mA

Full load V_{stab} = 6.5 V → 7.6 V
I = 43.8 mA → 53.5 mA

V_1	12.8	12.4	12.3	12.2	9.4	7.8
V_2	12.3	12.1	12.0	11.3	8.5	7.0
I (mA)	2.2	4.0	5.7	10.6	34.0	47.0

Practical Exercise 5m

Notable improvement from the results obtained in Practical Exercise 5c.

Practical Exercise 5n

Waveforms observed were as shown in Fig. 5.28 and Fig. 5.29.

D.6 Chapter 6. Transistors

Practical Exercise 6b

1.

	Link OUT	Link IN
V_{BE}	0	0.7 V
V_{CE}	12.0 V	0.1 V
I_C	0	0.2 A

2. The transistor can switch from one state to the other in less than 1 μs. As

the frequency of the signal increases towards 100 kHz the switching time increases towards 2 μs.

Practical Exercise 6c

1. The value of resistance required in the circuit of Fig. 6.8 to cause the lamp to come on when the ORP 12 is shielded from the light is 4.2 kΩ.
2. When R and the ORP 12 are interchanged the lamp comes on when light falls on the ORP 12.

Practical Exercise 6d

When the wander lead is touched from 0 V to the base of the ON transistor, this transistor is switched OFF and the other transistor is turned ON. If the wander lead is touched on this transistor base it switches OFF and the other transistor comes ON again.

Practical Exercise 6e

1. V_{in} = 100 mV peak-to-peak, V_{out} = 10 V peak-to-peak. Therefore,

$$\text{voltage gain} = \frac{10}{0.1} = 100$$

2. As the input signal amplitude is increased the output waveform becomes distorted, i.e., the top and bottom of the wave appear to be clipped.
3. With the bias set to give V_{CE} = 9 V, the top of the output waveform becomes clipped first; owing to the bias being set high the transistor reaches cut-off more easily.
4. With the bias set to give V_{CE} = 3 V, the bottom of the output waveform becomes clipped first, since the bias causes the transistor to be nearer to saturation.
5. The bandwidth is the range of frequencies over which the response of the amplifier is greater than half power (i.e., better than – 3 dB from its maximum response). f_1 = 300 Hz, f_2 = 400 kHz. Therefore,

$$\text{bandwidth} = f_2 - f_1 = 400 - 0.3 = 399.7 \text{ kHz}.$$

Practical Exercise 6f

The waveforms of input signal and output voltage are shown in Fig. D.5.

Practical Exercise 6g

Input AB = 4 V peak-to-peak. Output CB = 16 V peak-to-peak. Therefore,

$$\text{voltage gain} = \frac{16}{4} = 4$$

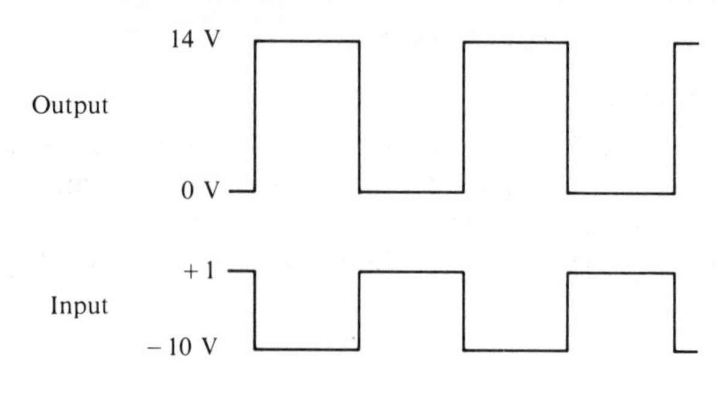

(*a*) Waveforms for Fig. 6.13(*a*)

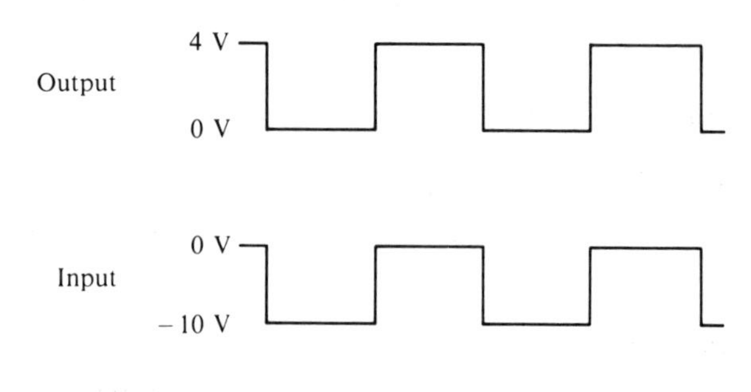

(*b*) Waveforms for Fig. 6.13(*b*)

Fig. D.5 FET as a switch.

When the FET gate lead was disconnected there was virtually no change in the amplitude of the input signal, thus indicating that the FET amplifier has a high input impedance.

D.7 Chapter 7. Unijunction transistor

Practical Exercise 7a

$V_{EB_1} = 6.7$ V. Therefore,

$$\eta = \frac{V_{EB_1}}{V_{BB}} = \frac{6.7}{10} = 0.67.$$

Practical Exercise 7b

The waveforms observed do compare with those shown in Fig. 7.3.

D.8 Chapter 8. Thyristors and triacs

Practical Exercise 8a

Using a d.c. supply, the lamp remains ON once the thyristor is triggered. The supply voltage must be switched off to stop the thyristor conducting.

With an a.c. supply, the lamp can be turned ON and OFF by adjustment of the resistance in the gate circuit. This is because the supply voltage is reversed each half-cycle.

Practical Exercise 8b

1. The thyristor can only be turned on by using a charged capacitor when the capacitor is 0.01 μF or greater, and the positive terminal is connected to A in the circuit of Fig. 8.7.
2. With the lamp ON, the thyristor was switched OFF by connecting a charged capacitor across the thyristor terminals CD. The minimum size of capacitor which switched the thyristor off was 0.47 μF.

D.9 Chapter 9. Pulse shaping

Practical Exercise 9a

The waveforms obtained were similar to those shown in Fig. 9.5(*b*) and (*c*), respectively.

Practical Exercise 9b

The waveforms obtained were similar to those shown in Fig. 9.9(*b*). When the diode was reversed, the output pulses were reversed to give a train of negative pulses only.

Practical Exercise 9c

The output waveform is a clipped sine wave, but since the input signal is large the output is virtually a rectangular waveform. Adjustment of the bias causes the output waveform symmetry above and below 5 V to be changed.

Practical Exercise 9d

The waveforms obtained were the same as those shown in Fig. 9.15. The modification shown in Fig. 9.16(*a*) caused a change in the mark-to-space ratio with some change in frequency, and the modification shown in Fig. 9.16(*b*) caused a change in frequency with some change in the mark-to-space ratio.

Practical Exercise 9f

The waveforms obtained for the monostable multivibrator are as shown in Fig. 9.19.

Practical Exercise 9g

The waveforms obtained for the Schmitt trigger were similar to those shown in Fig. 9.20(*b*).

Practical Exercise 9h

The waveforms obtained for the 741 Schmitt trigger circuit were similar to those shown in Fig. 9.20(*b*). Variation of the 25K0 caused a change in trigger levels.

D.10 Chapter 10. Amplifiers

Practical Exercise 10a

Applied voltage = 10 V peak-to-peak.
Voltage across 1K0 = 10 mV peak-to-peak.
Voltage across 100K0 = 1 V peak-to-peak.

The higher the load resistance the less damping (or loading) effect is observed on the input signal.

Practical Exercise 10b

With the common emitter amplifier, input AB = 50 mV peak-to-peak, output CB = 4 V peak-to-peak. When the transistor base lead was disconnected the input AB increased to 2 V peak-to-peak so that amplifier impedance was loading the circuit quite dramatically.

With the emitter follower amplifier, input AB = 1.6 V peak-to-peak, output CB = 1.5 V peak-to-peak. When the transistor base lead was disconnected there was virtually no change in the input AB so that this amplifier did not load the circuit at all.

Practical Exercise 10c

Amplifier gain = 1400

The gain falls as the load resistance is decreased.

The gain/frequency response is shown in curve (a) in Fig. D.6 without negative feedback.

When negative feedback is applied (and R_1 connected) the gain fell to 140. The gain remains reasonably constant as load resistance is varied (with n.f.b.). The gain/frequency response with feedback is shown in curve (b) in Fig. D.6.

Practical Exercise 10d

Amplifier gain = 35
With PQ connected and R_4 disconnected, amplitude = 300 mV at 750 Hz.
Phase of signal at X = 60°, Y = 120°, and P = 180°

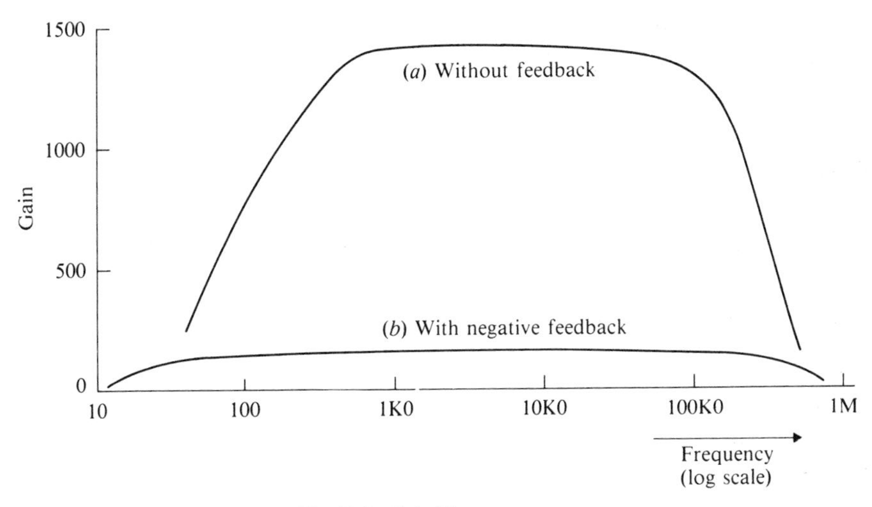

Fig. D.6 Gain/Frequency response.

Practical Exercise 10e

Range of output voltage (offset) as the offset adjust variable resistance is varied is ± 30 mV.

Practical Exercise 10f

1. $V_{in} = +2$ V, $V_{out} = -10$ V

$$\text{i.e., gain} = -\frac{10}{2} = -5 \text{ which compares with } \frac{100}{20} = 5$$

As V_{in} is increased negatively, the output *saturates* at +12.8 V, the input voltage under these conditions is about –2.55 V. As V_{in} is increased positively, at about 2.55 V the output *saturates* at about –12.8 V.

2. $V_{in} = 1$ V peak-peak, $V_{out} = 5$ V peak-peak.

$$\text{Voltage gain} = \frac{5}{1} = 5$$

but the phase is changed by 180°. Thus, gain = –5.

$$\text{this compares exactly with } -\frac{100\text{K0}}{20\text{K0}} = -5$$

With $R_2 = 20$K0, gain is reduced to unity and the amplifier becomes an *inverter* or *sign-changer.*

Practical Exercise 10g

1. $v_1 = 3$ V, $v_2 = 2$ V and $v_{out} = -5$ V
2. $v_1 = 5$ V peak-peak, $v_2 = 2$ V peak-peak, and $v_{out} = -10$ V peak-peak. At different frequencies, the output represents a '*mix*' of the input signals.

Practical Exercise 10h

The gain is unity for both d.c. and a.c. inputs. No phase change.

Practical Exercise 10i

1. $v_{in} = +2$ V, $v_{out} = +4$ V. Gain $= +\frac{4}{2} = 2$

   Compares with gain $= +\frac{(10 + 10)}{10} = \frac{20}{10} = 2$

2. $v_{in} = 3$ V peak-peak, $v_{out} = 6$ V peak-peak. Gain $= \frac{6}{3} = 2$ with no phase shift.

Practical Exercise 10j

$v_1 = 5$ V, $v_2 = 2$ V, $v_{out} = -3$ V

$$v_{out} = -20\left(\frac{5}{20} - \frac{2}{20}\right) = -(5 - 2) = -3\text{V, as measured.}$$

Practical Exercise 10k

1. *d.c.* $v_1 = 2$ V, v_{out} gradually increases (with opposite polarity to the input) until saturation is reached (–12.8 V in this case).
2. *a.c.* $v_1 = 2$ V peak-peak, square wave. v_{out} is a triangular waveform.

Practical Exercise 10l

The output waveform is observed to consist of narrow pulses as the input signal changes polarity.

Practical Exercise 10m

A square wave is observed at A, and a triangular wave at B. Frequency is adjustable using the 10K0 variable resistor.

Practical Exercise 10n

Satisfactory results were obtained, which would enable this arrangement to be suitable for a pocket radio.

D.11 Chapter 11. Optoelectronics

Practical Exercise 11a

With maximum resistance in circuit, the emitted light from the LED was hardly visible. Under these conditions $V = 1.75$ V and $I = 0.1$ mA. With minimum resistance in circuit, the LED was reasonably bright and $V = 2.48$ V, $I = 7$ mA. Therefore, a useful indicator can be made by using an LED with a series resistor of between 470 Ω and 2 kΩ.

Practical Exercise 11b

The output pulses observed were similar to those shown in Fig. 11.20. The frequency of occurrence of pulses is proportional to speed, i.e., for 1800 rev min^{-1}:

$$\text{speed} = \frac{1800}{60} = 30 \text{ rev s}^{-1}$$

and with 10 holes in the disc, 10 pulses are produced each revolution. Hence,

$$\text{frequency of pulses} = 30 \times 10 = 300 \text{ pulses per second (Hz)}$$

and so

$$\text{periodic time} = \frac{1}{300} = 0.00333 \text{ s.}$$

Therefore,

$$\text{time between pulses on CRO} = 3.33 \text{ ms.}$$

for a speed of 1800 rev min^{-1}.

D.12 Chapter 12. Logic and Digital Electronics

Practical Exercise 12a

The truth tables for the AND and OR gates were as shown in Fig. 12.1(*c*) and Fig. 12.2(*c*), respectively. The actual logic 'low' ('0') state for the AND gate was +0.7 V (consistent with the voltage drop across a forward biased silicon p n junction), and the 'high' state was +6 V. When the gate was loaded with a 6 V lamp, the output F decreased to zero (i.e., the lamp did not light).

The 'high' level with the OR gate was 5.3 V and the 'low' level was 0 V. When loaded with a 6 V lamp, the lamp lit with virtually no change in output F. This is due to the fact that in the 'high' state a diode and the lamp are in series across the supply.

Practical Exercise 12b

The truth table for the RTL gate is as shown below:

A	B	F	O/P voltage at F	
0	0	1	24 V	
0	1	0	0.05 V	Transistor hard on
1	0	0	0.05 V	
1	1	0	0.05 V	

The logic '1. is therefore *ideal*, but the logic '0' is slightly higher than the ideal of 0 V.

Input current to switch gate ≈ 2.3 mA.

Each similar gate therefore causes the high output of the driving gate to fall by approximately 2.3 V. Therefore the fan-out is less than 3. With approximately 75K0 input resistance, the drive current to saturate the transistor is reduced to about 0.3 mA. Therefore, the fan-out is increased to about 20.

Practical Exercise 12c

The truth table for the DTL gate is as shown below:

A	B	F	O/P voltage at F
0	0	1	5 V
0	1	1	5 V
1	0	1	5 V
1	1	0	0.07 V

Practical Exercise 12d

The truth table for the SN 7400 TTL NAND gate is as shown below:

A	B	F	O/P voltage at F
0	0	1	3.7 V
0	1	1	3.7 V
1	0	1	3.7 V
1	1	0	0.1 V

The AND and OR functions, using NAND gates only were verified.

Practical Exercise 12e

Sink Current = 1.1 mA

Practical Exercise 12f

Source Current = 6 μA

Practical Exercise 12g

Truth table:

A	B	F	X	Y
0	0	0	0	0
0	1	1	0	1
1	0	1	1	0
1	1	0	0	0

Thus, point X gives an output '1' when $A > B$, and point Y gives an output '1' when $A < B$.

Practical Exercise 12h

The display indicates that of a four-bit binary counter. When the R_0 is connected to '1' the display is returned to zero, i.e., 0000, and when the R_0 is connected to '0' the count proceeds as before.

When the arrangement is set up as shown in Fig. 12.24, the display is that of a decade counter going through the sequence 0 through to 9 as pulses are applied. The display is *reset* to *zero* when R_0 is set to logic '1', and continues counting when R_0 is set back to logic '0'.

Practical Exercise 12i

The circuit arrangement shown in Fig. 12.25 is that of a four-bit binary counter with the addition of a logic decoding network. In this case, the decoder detects when Q_B and Q_C are at logic 1 at the same time. This corresponds to a denary count of 6. The output of the decoder is used to reset the display to zero. Thus, the highest number displayed is *one* previous to the decoded six, i.e., 5, so that this system counts repeatedly in the sequence 0, 1, 2, 3, 4, 5, 0, 1, etc.

D.13 Chapter 13. Industrial Control

Practical Exercise 13a

Full-wave control is provided for the lamps, one thyristor conducting during each half-cycle of the supply. The waveforms are similar to those shown in Fig. 13.3(b).

Practical Exercise 13b

Acceptable speed control was achieved with this arrangement by adjustment of the 25K0 variable resistor. The waveform across the motor armature was similar to the top waveform in Fig. 13.3(*c*) and the addition of the flywheel diode eliminated the negative spike on the waveform.

Practical Exercise 13c

The transistor chopper arrangement provides a useful form of speed control, over a reasonable range for the motor used.

Practical Exercise 13d

In this system, the astable multivibrator outputs are fed to the gates of the thyristors so that the conduction angle can be varied by adjustment of the frequency. Here again, a useful range of speed control was achieved.

D.14 Chapter 14. Electronic measurements and Fault Diagnosis

Practical Exercise 14a

With sine wave supply:

1. CRO peak-to-peak voltage = 10 V. Therefore,

$$\text{r.m.s.} = \frac{10}{2} \times 0.707 = 3.54 \text{ V.}$$

2. DMM voltage = 3.55 V.
3. AVO Multimeter Model 8 voltage = 3.25 V.

With square wave supply:

4. CRO = 10 V peak-to-peak = 5 V r.m.s.
5. DMM = 4.5 V. Therefore,

   r.m.s. reading = 4.5 x 1.11 = 5.0 V,

   since the DMM has been calibrated to read r.m.s. values on a *sine* wave.
6. AVO = 4.1 V. Therefore,

$$\text{r.m.s. reading} = 4.1 \times 1.11 = 4.6 \text{ V.}$$

From the results it can be seen that the CRO gives quite accurate results, although it is more difficult to read and interpret. The DMM again gives the most accurate results, but it is necessary to modify the reading on waveforms other than sine wave. The AVO Multimeter introduces a loading error on a.c. ranges and, therefore, the readings are in error by less than 10 per cent in this case.

Practical Exercise 14b

The displays obtained are comparable to those shown in Fig. 14.5. The angle of the trace is changed as the 25K0 variable resistor is varied.

Practical Exercise 14c

The displays obtained are comparable to those shown in Fig. 14.6.

Practical Exercise 14d

The display obtained is the same as that shown in Fig. 14.7(*b*). As the frequency range is limited, the response changes to that shown in Fig. 14.7(*c*). The maximum response was varied between 8 kHz and 19 kHz as the inductive core was varied. When the 10K0 was connected across the tuned circuit, the amplitude of the display was reduced and the maximum response occurred at 7 kHz instead of 8 kHz.

Practical Exercise 14e

Amplifier input = 5 mV pk-pk; Amplifier output = 560 mV pk-pk

Therefore amplifier gain $= \frac{560}{5} = 112$

1. Input impedance, $Z_{in} = 5\ \text{k}\Omega$
2. Output impedance, $Z_{out} = 525\ \text{k}\Omega$

Practical Exercise 14f

I_B (mA)	0.03	0.06	0.09	0.12	0.15
I_C (mA)	3.4	7.0	11.1	14.6	18.1

The graph of these results is shown in Fig. D.7.

From the graph, the value of $h_{FE} = \frac{10.4}{0.086} = 121$

Practical Exercise 14g

When transistor is switched HARD ON, $V_{CE} = 46$ mV (i.e., V_{CEsat}). Under these conditions, $V_{BE} = 0.85$ V

Practical Exercise 14h

4. V_{CE} decreases with time, e.g., to 1.6 V in about 10s.
6. Blowing on the transistor slows down the rate of decrease.

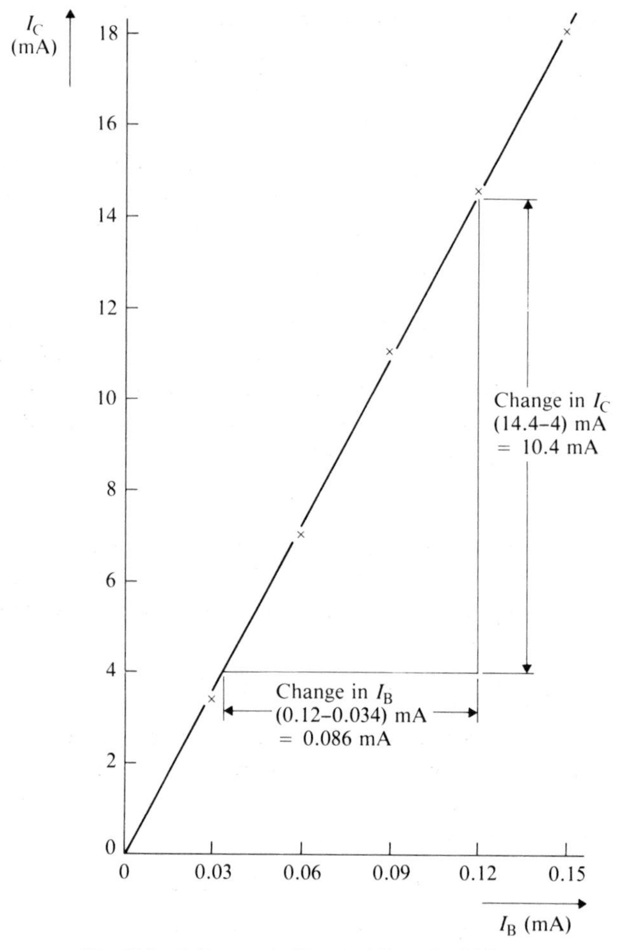

Fig. D.7 I_B/I_C **graph (Practical Exercise 14f).**

8. Heat applied to the transistor causes V_{CE} to decrease.
9. V_{CE} decreases, due to the increased leakage current adding to I_C.
11. V_{CE} decreases when warm fingers are placed in contact with the transistor.

Practical Exercise 14i

The fall in V_{CE} is much less noticeable than in Practical Exercise 14h.

Practical Exercise 14j

Typical results together with suggested outlines of diagnosis for each of the fault conditions (14j (2) – 14j (12)) are given below:

Normal conditions: $V_1 = 2.2$ V; $V_2 = 4.55$ V; $V_3 = 1.6$ V.

14j (2) $V_1 = 0.7$ V; $V_2 = 0.03$ V; $V_3 = 0$ V

With $V_2 = 0.03$ V, the transistor is switched hard ON. Base voltage is 0.7 V above the emitter which is at 0 V. The emitter must therefore be connected to the 0 V rail. Suspect component is C_E *short-circuited.*

14j (3) $V_1 = 0.8$ V; $V_2 = 8.4$ V; $V_3 = 0.2$ V

A higher than normal V_2 suggests that the transistor is not conducting as heavily as normal. This is consistent with a reduced I_E and a reduced V_3 (0.2 V in this case). The base voltage is sufficiently positive just to cause the transistor to conduct – but insufficient to create 'normal' conditions. A possible cause of these conditions is R_2 – *low resistance.*

14j (4) $V_1 = 3.0$ V; $V_2 = 3.0$ V; $V_3 = 2.36$ V

Since V_1 and V_2 are approximately the same, it is assumed that the transistor *collector/base junction is short-circuited.* Under these conditions, the voltage difference between V_2 and V_3 should be about 0.7 V – the forward voltage drop across a diode (b/e junction).

14j (5) $V_1 = 3.15$ V; $V_2 = 2.57$ V; $V_3 = 2.42$ V

The base voltage V_1 has increased, which causes an increase in base current, which in turn causes more collector current (and therefore emitter current). Thus V_2 falls, and V_3 increases. A possible cause for the increase in V_1 and thus I_B is that R_2 *is open-circuit.*

14j (6) $V_1 = 0.1$ V; $V_2 = 9.0$ V; $V_3 = 0.1$ V

Voltages V_1 and V_2 are equal, suggesting the *transistor b/e junction is short-circuited.* The transistor is cut-off, so that $V_2 = 9.0$ V.

14j (7) $V_1 = 0.75$ V; $V_2 = 9.0$ V; $V_3 = 0.1$ V.

With $V_2 = 9.0$ V, $I_C = 0$ mA. But, since the base voltage is about 0.7 V higher than the emitter, the transistor should be conducting. The 0.1 V at the emitter is consistent with the base current *only* flowing in R_E. It is therefore diagnosed that the *transistor c/b is open-circuit.*

14j (8) $V_1 = 0.75$ V; $V_2 = 0.1$ V; $V_3 = 0.1$ V.

Initially, it might be assumed that the c/e are short-circuited, since V_2 and V_3 are the same. However, this cannot be the case, because R_C and R_E would form a potential divider across the supply. With V_1 about 0.7 V higher than the emitter it is assumed that the bias circuit is satisfactory. The diagnosis is therefore that R_C *is open-circuit.* The voltage V_3 (0.1 V) may be accounted for by the connection of the voltmeter to the circuit.

14j (9) $V_1 = 2.5$ V; $V_2 = 9.0$ V; $V_3 = 0$ V.

Collector voltage = + V_{CC}, and emitter voltage = 0 V indicate that the transistor is cut-off. But, the potential at V_1 is slightly greater than normal – which suggests that $I_B = 0$. Therefore, it is assumed that the *transistor b/e junction is open-circuit.*

14j (10) $V_1 = 2.45$ V; $V_2 = 9.0$ V; $V_3 = 2.05$ V.

Since V_1 is slightly higher than normal, it is likely that $I_B = 0$; in which case I_C will be zero – this is consistent with $V_2 = 9.0$ V. It is therefore assumed that R_E *is open-circuit.* The voltage at V_3 can be accounted for by C_E being charged through the b/e junction – and this is the voltage V_3.

14j (11) $V_1 = 2.27$ V; $V_2 = 9.0$ V; $V_3 = 1.65$ V.

Since V_1 and V_3 are almost the same as for the normal conditions, it is assumed that I_B and I_C are normal. It is therefore assumed that R_C ***has a reduced resistance value.***

14j (12) $V_1 = 3.1$ V; $V_2 = 2.5$ V; $V_3 = 2.4$ V.

These results are very similar to those of 14j (5), and further circuit tests must be carried out to confirm the diagnosis. V_1 has increased, thus I_B increases causing an increase in I_C. Therefore V_2 decreases and V_3 increases. A possible cause for this fault could be that R_1 ***has a reduced resistance.***

Appendix E. Summary of Basic Theory

E.1 Materials

1. ***Conductors.*** Copper and aluminium for wiring. Brass for switch and plug parts, etc. Steel for conduit, etc. Nickel for plating components. Nichrome for heating wires. Silver and platinum for contacts. Lead/tin alloys for solder.
2. ***Semiconductors.*** Silicon and germanium for transistors and other modern electronic devices. Most industrial electronic devices are based on silicon.
3. ***Insulators.*** PVC and rubber for wire insulation. Paper and polythene for wire wrapping. Other plastics are used for insulation, certain oxides and Perspex.

E.2 Units

1. ***Current.*** Symbol I. Flows from *positive* to *negative* (+ve to −ve).
   Measured in *amperes* (symbol A).
   Small values of current may be measured in *milliamperes* (mA) – thousandths of an ampere – or *microamperes* (μA) – millionths of an ampere.
   Measured with an *ammeter*, which should have a *low* resistance (typically a fraction of an ohm).
2. ***Electromotive force, e.m.f.*** Symbol E. Cells, batteries, generators, and other sources of electricity are said to produce an e.m.f. measured in *volts* (symbol V).
3. ***Voltage or potential difference.*** Symbol V. This appears across any device through which current is flowing.
   Measured in *volts* (V).
   Large values of voltage may be measured in kilovolts (kV) – thousands of volts, e.g., 132 kV is the same as 132000 V.
   Measured with a *voltmeter*, which should have a very high resistance (typically 20000 Ω V^{-1}).
4. ***Resistance.*** Symbol R. This is the opposition to the flow of current.
   Measured in *ohms* (Symbol Ω – Greek capital 'omega').
   The circuit symbols commonly used are as shown in Fig. E.1.
   Measured with an ohmmeter.

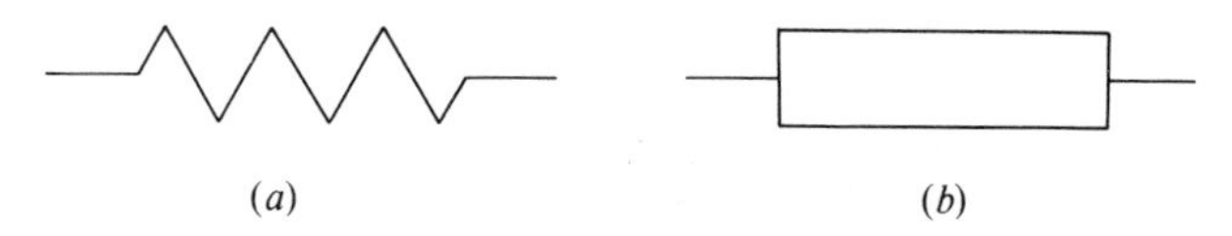

Fig. E.1 Circuit symbols for resistance.

5. ***Ohm's law.*** This relates current in amperes, voltage in volts, and resistance in ohms:

$$I = V/R, \qquad R = V/I, \qquad V = IR.$$

This really means that if the voltage across a resistance is doubled then the current through that resistance must have been doubled.

Assuming a constant supply voltage, the greater the resistance in a circuit the lower the curent.

E.3 Use of meters

1. ***Ammeter.*** To measure current *break* the circuit and insert the ammeter in the gap. *Positive* (generally *red*) terminal of the ammeter to the positive supply side.
2. ***Voltmeter.*** To measure voltages or p.d.'s fit prods to the meter leads and prod *across* the two points between which the unknown voltage appears. There is *no* need to break the circuit. If there is no resistance between the points and current is flowing then the voltmeter will not show a reading, i.e., across a closed switch which is making good contact the p.d. will be negligible.
3. ***Ohmmeter.*** Short the leads together, adjust the meter to read full scale, i.e., zero ohms. Connect unknown resistor between meter lead ends and read resistance scale from right to left on meter.

E.4 Power

In general, where there is a current flowing and a p.d. exists, then electrical power is being dissipated. This is normally transformed into some other form of energy such as heat, light or mechanical.

Electrical power is measured in *watts* (symbol W). Larger powers are measured in kilowatts (kW). 1 kW is equivalent to 1000 W.

Energy is measured in watt hours or kilowatt hours.

$$\text{Power } (P \text{ watts}) = \text{Voltage } (V \text{ volts}) \times \text{Current } (I \text{ amperes}),$$

$$P = V \times I$$

Thus, the 1 kW electric fire shown in Fig. E.2 will allow a current of 4 A to

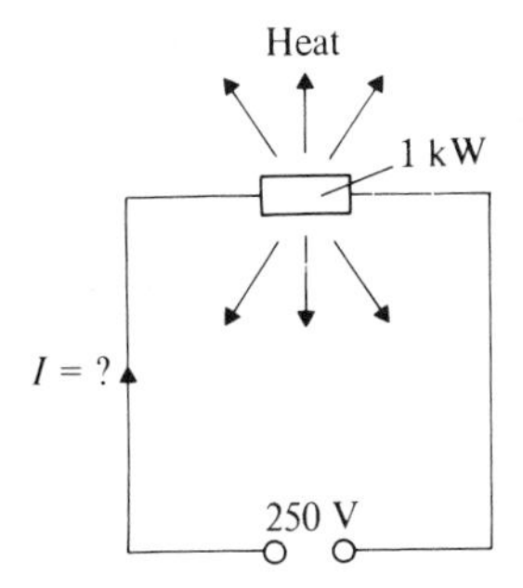

Fig. E.2 Electric fire circuit.

flow through it. This has been calculated by using the formula:

$$P = V \times I \quad \text{or} \quad I = P/V$$

In this case, the voltage $V = 250$ V, and so

$$I = \frac{1000}{250} = 4 \text{ A}$$

EXAMPLE E.1

The applied voltage to an electric fire element is 250 V and the current flowing is 5 A. Calculate the resistance and wattage of the element.

Solution

$$R = \frac{V}{I} = \frac{250}{5} = 50\ \Omega.$$

Hence,

$$P = V \times I = 250 \times 5 = 1250 \text{ W}.$$

Therefore,

$$P = 1.25 \text{ kW}.$$

E.5 Resistance in series and parallel

1. ***Series.*** This is very simple, just add them up. A 4 Ω resistor in series with a 5 Ω resistor in series with a 7 Ω resistor would be replaced by a 4 + 5 + 7 = 16 Ω resistor.

2. ***Parallel.*** Use the formula

$$\frac{1}{R} = \frac{1}{R_1} + \frac{1}{R_2} + \frac{1}{R_3} + \ldots$$

Thus, the equivalent resistance for the three parallel connected resistors in Fig. E.3 may be obtained as follows:

$$\frac{1}{R} = \frac{1}{4} + \frac{1}{5} + \frac{1}{7}.$$

Hence,

$$\frac{1}{R} = \frac{35 + 28 + 20}{140} = \frac{83}{140}.$$

Therefore,

$$R = \frac{140}{83} = 1.69\ \Omega.$$

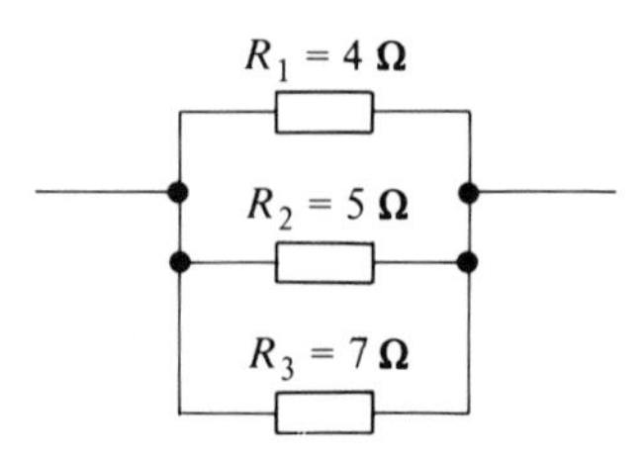

Fig. E.3 Resistors in parallel.

Note that if we have two *equal* resistors in parallel, then the current will split equally between them, and the equivalent resistance is *half* the value of one resistor.

If there are only *two* resistors in parallel, then:

$$\frac{1}{R} = \frac{1}{R_1} + \frac{1}{R_2} = \frac{R_2 + R_1}{R_1 R_2}.$$

Therefore,

$$R = \frac{R_1 R_2}{R_1 + R_2} = \frac{\text{product}}{\text{sum}}.$$

Your answer for parallel resistors must always be less than the smallest resistor.

E.6 Variable resistors

The commonly used circuit symbols for variable resistors are as shown in Fig. E.4.

Fig. E.4 Circuit symbols for variable resistance.

Variable resistors will control the current flowing in a circuit; e.g., the dimming of the instrument lights on a car dashboard is achieved by turning the knob of a variable resistance, as shown in Fig. E.5. Variable resistors are also used in some fuel gauges in cars.

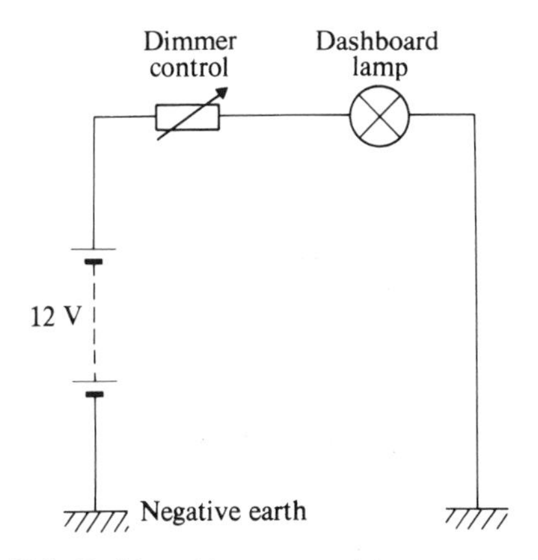

Fig. E.5 Dashboard instrument light dimming circuit.

Appendix F. Component List

F.1 Components supplied with basic electronic learning kit

10K0	560K0	1M0			
8K2	470K0	470 μF 50 V		ORP 12	
4K7	330K0	100 μF 16 V		LED	
2K2	150K0	10 μF 16 V		FIL.LAMP. 12 V 1 W	
1K0	100K0	10 μF 16 V	TRIAC	FIL.LAMP. 12 V 1 W	
1K0	100K0	4.7 μF 50 V	103YY (THY)	DIAC.	
680R	47K0	1 μF	2N 3819	10V0ZD	
470R	47K0	0.47 μF	TIS 43 (UJT)	4V7 ZD	
240R	20K0	0.1 μF	2N 3053	1N 4001	
180R	20K0	0.047 μF	2N 3053	1N 4001	
100R	15K0	0.01 μF	BC 461	1N 4001	
10R	10K0	0.01 μF	ZTX 300	1N 4001	10 LINKS

F.2 Supplementary component list

(Available from Beal-Davis Electronics Ltd, Newtown Road, Worcester)

	47 R		ZTX 300
	100R	2	741
	200R		ZN 414
	240R		LM 380
	1K0		Loudspeaker (8 Ω)
	1K5		Coil-var. tuning.
	2K2		OPB 120
	4K3		103 YY
2	4K7		7-seg. LED
	5K6		SN 7400 logic chips
	6K8		U2 Cell
	10K0		0–100 μA movement.
2	20K0		
	100K0		Choke
	180K0		± 15 V supply
	390K0		Breadboard (for Logic)
	470K0		
	1M0		
2	10K0Pot		
	50K0Pot		
	100K0Pot		

Thermistor GL 23
0.1 μF
1 μF
10 μF
Var. Cap 500pF

Neon (1650 SB Farnell) 60 V

Index